Michael Tooma on

Managing Asbestos in Workplaces

Disclaimer

Wolters Kluwer

Michael Tooma on
Managing Asbestos in Workplaces

Michael Tooma

CCH AUSTRALIA LIMITED
GPO Box 4072, Sydney, NSW 2001
Head Office Macquarie Park
Phone: (02) 9857 1300
Customer Support
Phone: 1 300 300 224
www.wolterskluwer.cch.com.au
Book Code: 10078258-0001

About Wolters Kluwer

Wolters Kluwer is a leading provider of accurate, authoritative and timely information services for professionals across the globe. We create value by combining information, deep expertise, and technology to provide our customers with solutions that contribute to the quality and effectiveness of their services. Professionals turn to us when they need actionable information to better serve their clients.

With the integrity and accuracy of over 45 years' experience in Australia and New Zealand, and over 175 years internationally, Wolters Kluwer is lifting the standard in software, knowledge, tools and education.

Wolters Kluwer – *When you have to be right*.

Enquiries are welcome on 1300 300 224.

"Cataloguing-in-Publication Data available through the National Library of Australia."

First edition 2020

ISBN 978-1-925894-47-9

Contents

Preface

Much of what has been written on Work Health and Safety law has understandably focused on the duty of care. The audience for these books has been primarily safety practitioners and lawyers who practice in the space. My Due Diligence series sought to venture away from that target audience, targeting senior managers instead. While the feedback from that series was exceptionally positive, a gap still remained in relation to line managers who are left to interpret the duty of care in practice. This series is aimed at that audience.

It is intended as a companion series to my Due Diligence series aimed at line managers. In that regard it is not focused on a discussion of general principles such as the duty of care or the due diligence duty or even the industrial manslaughter obligations recently introduced in Queensland and foreshadowed in other jurisdictions. Instead, the book focuses on specific obligations and practical strategies for dealing with critical risks.

For a line manager, a practical discussion of critical risks, developed based on case examples, is a far more beneficial reference than a broad discussion of principles. With that in mind, each chapter of the books in this series begins with an illustrative case example. The chapters then develop the practical strategies and recommendations for the issue being discussed. In that context, any requirements of the Work Health and Safety Regulations are discussed and explained. The books also include the guidance contained in relevant Codes of Practice to ensure that any recommended approach is comprehensive and considered.

The series adopts my plain English style of writing. Diagrams are used to illustrate concepts. Chapters and book lengths have been deliberately kept to a minimum to maximise accessibility.

Michael Tooma
February 2020

Wolters Kluwer Acknowledgments

Wolters Kluwer wishes to thank the following who contributed to and supported this publication:

Director, General Manager – Research and Learning: Lauren Ma

Associate Director, Regional Head of Content: Diana Winfield

Head of Legal Content: Carol Louw

Books Coordinator: Jackie White

Editor: Geeta Shyam

Cover Designer: Anjali Kakkad

About the Author

Michael Tooma is Australian Managing Partner of global firm Clyde & Co. He has specialised in work health and safety law for over 20 years and is the author of over 25 books on work health and safety law.

Michael has taught work health and safety law at several Australian and international tertiary institutions including the University of New South Wales, University of Sydney and Auckland University.

Michael is consistently ranked as a leading work health and safety lawyers by legal directories. In 2017, Legal 500 Asia Pacific quoted client testimonials that called him "phenomenal" and in 2018 that he is "engaging and knowledgeable" and "he can answer any question posed to him". Chambers 2016 said that Michael "has a top reputation in Australian work health and safety practice, with clients saying he is 'an absolute authority' on safety legislation who can 'talk about it in applicable layman's terms'". Best Lawyers has ranked him as one of Australia's best work health and safety lawyers since the inception of that category in 2008.

About the Author

Michael Tooma is Australian Managing Partner of global firm Clyde & Co. He has specialised in work health and safety law for over 20 years and is the author of over 25 books on work health and safety law.

Michael has taught work health and safety law at several Australian and international tertiary institutions including the University of New South Wales, University of Sydney and Auckland University.

Michael is consistently ranked as a leading work health and safety lawyers by legal directories. In 2017, Legal 500 Asia Pacific quoted client testimonials that called him "phenomenal" and in 2018 that he is "engaging and knowledgeable" and "he can answer any question posed to him". Chambers 2016 said that Michael "has a top reputation in Australian work health and safety practice, with clients saying he is 'an absolute authority' on safety legislation who can 'talk about it in applicable layman's terms'". Best Lawyers has ranked him as one of Australia's best work health and safety lawyers since the inception of that category in 2008.

CHAPTER 1

INTRODUCTION

Refurbishment of kindergarten results in asbestos prosecution

A construction company was prosecuted and fined $4,500 for two breaches of the *Work Health and Safety Regulations* relating to asbestos management.[1]

The company was engaged to undertake refurbishment works at a kindergarten commencing in the school holidays of September 2015. The renovation works required removal and disposal of existing ceiling tiles, vinyl flooring and external awning, and removal of the internal kitchen and cabinets for reuse. Asbestos was not identified at any time, other than in the internal wall sheeting which was not being removed.

The building housing the kindergarten was believed to have been constructed in the mid-1970s due to the type of building fabric used. Plasterboard ceiling tiles were installed throughout the building.

In approximately 2005 the plasterboard ceiling tiles over the kitchen, store, staff and student toilets were removed and replaced by fibrous cement tiles that were presumed not to have comprised asbestos containing material, for the reason that asbestos had not been used in building products in Australia since the 1980s, with the use of all asbestos products being finally banned in Australia on 31 December 2003.

An asbestos audit was conducted of the building on 7 November 2013 and minor repair works were undertaken in early 2015.

In May 2015, a firm of architects were engaged to act as project managers for the kindergarten's refurbishment. The company was engaged through the project manager to undertake the refurbishment works.

Three workers were engaged to perform the removal work which was undertaken between 19 and 21 September 2015. In the course of the removal work ceiling tiles were smashed and vinyl tiles which were glued to the floor were removed with a scraper.

---

1 *Work Health and Safety Queensland v E218806*, Cairns Magistrate's Court, 28 March 2018, Magistrate Pearson.

No protective equipment was worn by the workers and the demolished material was simply transported to the local dump without being wrapped or treated in any way.

A council employee advised the workers on 21 September 2015 that asbestos was suspected of being present in the materials which had been transported to the dump. From then, the remaining materials removed were wrapped in plastic and left on site for removal by council workers.

The site was not properly decontaminated, with asbestos dust on the floor being swept up, and the verandah hosed.

None of the workers' clothing, vehicles or tools were decontaminated.

Subsequent testing of sample taken from the kindergarten and dump sites conducted by a NATA-accredited laboratory revealed that chrysotile asbestos was present at both sites.

On 28 March 2018, the defendant pleaded guilty in the Cairns Magistrates Court to breaching s 422(1) and 448 of the *Work Health and Safety Regulation 2011*.

Magistrate Sandra Pearson fined the defendant $4,500 and ordered professional and court costs totalling $1,692.30. After the incident, the company had contributed $35,000 towards the cost of site decontamination. The court ordered that no conviction be recorded.

¶1-010 What is asbestos?

Asbestos is a naturally occurring mineral fibre that sadly has been mined and used extensively around the world for many years. As the Australian Asbestos Safety and Eradiation Agency observes:

"Asbestos was once considered to be a very useful mineral because it is flexible, strong, affordable and can insulate from heat and electricity. Because of this, it was commonly used in the construction of home and building".[2]

Indeed, Australia was one of the most prolific users of asbestos in the world. Consistent with its use globally, it was used in roofing, thermal and electrical insulation, cement pipe and sheets, flooring, gaskets, friction materials such as brake pads, coatings and compounds, plastics, textiles, paper, mastics, thread, fibre jointing and milliboard.[3] Asbestos use was only banned in Australia on 31 December 2003.

Homeowners and businesses in Australia now live with the legacy of those bad decisions. Approximately one-third of all homes in Australia contain asbestos products.[4]

Asbestos is carcinogenic to humans.[5] Exposure to asbestos, including chrysotile, causes cancer of the lung, larynx and ovaries, and also mesothelioma (a cancer of the pleural

2 Asbestos Safety and Eradication Agency, "About Asbestos", accessed online at www.asbestossafety.gov.au October 2019.

3 International Agency for Research on Cancer (IARC) (2012), "Asbestos (Chrysotile, Amosite, Crocidolite, Tremolite, Actinolite and Anthophyllite)", pp in IARC (2012), Arsenic, Metals, Fibres, and Dusts, IARC Monographs on the Evaluation of Carcinogenic Risks to Humans Volume 100C, Lyon, France, 2012, pp 219–234 at 221.

4 ASEA (2019), op cit 1.

5 IARC (2012), op cit 2, at p 294.

and peritoneal linings). Asbestos exposure is also responsible for other diseases such as asbestosis (fibrosis of the lungs), plaques, and thickening and effusion in the pleura.[6]

Currently, about 125 million people in the world are exposed to asbestos at the workplace. Approximately half of the deaths from occupational cancer are estimated to be caused by asbestos. In addition, it is estimated that several thousand deaths annually can be attributed to exposure to asbestos in the home.[7]

¶1-020 Asbestos in consumer products

According to the Asbestos Eradication Agency, in 2016–17 there were a number of reported asbestos-containing materials detected by the Australian customs border.

In August 2016, friable asbestos was detected in new plant equipment imported from China to be installed at the Port Pirie smelter redevelopment in South Australia.[8]

In June 2017, the Heads of Workplace Safety Authorities issued a consumer and retailer alert regarding the sale of children's crayons within Australia that were found to contain trace amounts of asbestos. Asbestos brake shoes were also detected in Vespa motor scooters and electrical scooters imported into Australia. It was also detected in a range of asbestos parts found in Polaris youth quad bikes resulting in a nationwide recall of several models.[9]

¶1-030 Asbestos exposure in the community

The Australian Government has created a register to record the details of members of the community who think that they may have been exposed to asbestos. The register is called the National Asbestos Exposure Register (NAER). Table 1 below sets out the NAER per jurisdiction for 2016–17.

As Table 1 indicates, 1,770 people have reported possible exposure to asbestos in Australia in the 2017 financial year, 1,560 people in 2016 and 1,400 in 2015.[10] Clearly

Table 1.1: NAER Register (Adapted from www.asbetossafety.gov.au)

Financial Year	ACT	NSW	Qld	Vic	Tas	SA	WA	NT	Total
2016–2017	86	478	279	161	92	122	491	61	**1770**
2015–2016	192	537	145	199	41	111	310	25	**1560**
2014–2015	383	369	123	181	65	79	121	79	**1400**

6 World Health Organization, "Asbestos: Elimination of asbestos-related diseases", 15 February 2018, accessed online at www.who.int October 2019.

7 Ibid.

8 Asbestos Eradication Agency, "Importation of asbestos containing material into Australia", accessed online at www.asbestossafety.gov.au

9 Ibid.

10 Ibid.

that is an unacceptable level of exposure. Given the link between asbestos exposure and asbestos-related terminal illnesses, we are likely to be dealing with the health consequences of these exposures for many years to come.

¶1-040 Scope of the book

This book is concerned with the obligations of persons conducting a business or undertaking (PCBU) in relation to asbestos. Specifically, it is written from the perspective of a PCBU who has control or management of a workplace which contains asbestos. This may be by virtue of ownership of premises used as workplaces — such as is the case for landlords. It may be by virtue of occupying such premises — such as is the case for tenants. While the book outlines the obligations of licensed asbestos removal contractors, it does so to outline for the PCBU who commissions the asbestos removal work what to expect from such contractors so that they can adequately manage the engagement. In that regard, the book does not address the extensive licensing requirements set out in the legislation.

The focus of the book is on proactively managing asbestos risks in the workplace. Consistent with the objective of the legislation and risk management principles, this begins with elimination or eradication of asbestos from workplaces. If asbestos is to remain in the workplace because it is not reasonably practicable to eliminate it, then there are a number of obligations that apply to PCBUs with management or control of the workplace in relation to managing the risks arising from the presence of asbestos.

An asbestos finding at a workplace is a serious event. If workers have been exposed to airborne asbestos fibres, it will also be a highly, emotionally charged event. Promptly and comprehensively responding to such incidents is crucial not only from the perspective of minimising the health impact from that incident but also the psychological effect of it.

CHAPTER 2

LEGAL FRAMEWORK

Bridge repair results in asbestos prosecution

A major provider of transport infrastructure projects in Queensland responsible for building and maintaining roads and bridges has been prosecuted and fined $175,000 in relation to an asbestos-related breach of its duty of care.[1]

Repair works were undertaken to a bridge over 5 months in 2012. The bridge contained asbestos, yet there was no documentation indicating this and no safety procedures outlining asbestos precautions.

The workers were not informed that the bridge contained asbestos and they were required to use power tools for a period of two weeks, which generated significant dust. They were not provided with appropriate personal protective equipment (PPE).

The defendant pleaded not guilty in the Ipswich Magistrates Court to breaching s 32 of the *Work Health and Safety Act 2011*. Following a four-day trial, it was found guilty on 1 June 2017 of having failed to meet its work health and safety duties and was sentenced on 24 July 2017.

Magistrate Virginia Sturgess fined the defendant $175,000 and costs. No conviction was recorded.

In reaching a decision, the magistrate noted that there were no comparable sentences to provide guidance on the matter. This was a serious offence and there was a significant failure at the hands of the defendant. There was no identification of the asbestos, no testing, no safety plan and no mention or consideration of the presence of asbestos.

1 *Work Health and Safety Queensland v E165293*, Ipswich Magistrates Court, 24 July 2017, Magistrate Virginia Sturgess.

> Magistrate Sturgess found that at least six workers were exposed to respirable asbestos fibres and these six workers now had a lifetime of wondering if they will develop an asbestos-related disease.

Magistrate Sturgess' observation in relation to the lasting psychological trauma associated with asbestos exposure is one of the many reasons why asbestos incidents are very sensitive and require a thoughtful and compassionate approach. Not only does exposure to asbestos fibres present a significant risk of contraction of a number of terminal asbestos-related illnesses, the latent onset of those illnesses adds to the trauma associated with the exposure. The relevant worker is left with a lifetime of wondering whether they will contract the illness.

¶2-010 Definition of asbestos

Asbestos is defined in cl 5 of the *Work Health and Safety Regulations*[2] (WHS Regulations) as "the asbestiform varieties of mineral silicates belonging to the serpentine or amphibole groups of rock forming minerals including the following:

(a) actinolite asbestos,

(b) grunerite (or amosite) asbestos (brown),

(c) anthophyllite asbestos,

(d) chrysotile asbestos (white),

(e) crocidolite asbestos (blue),

(f) tremolite asbestos,

(g) a mixture that contains 1 or more of the minerals referred to in paragraphs (a) to (f)".

Asbestos-containing material (ACM) is a phrase commonly used in the WHS Regulations. It is defined to mean "any material or thing that, as part of its design, contains asbestos".

¶2-020 Duty of care

The primary source of legal obligations in relation to asbestos arises from the duty of care owed to workers and others at a workplace.

A person conducting a business or undertaking (PCBU) has a duty to ensure, so far as is reasonably practicable, the health and safety of three categories of workers at work: (a) workers directly engaged by them, (b) workers caused to be engaged by them and (c) workers whose activities they can influence or direct.[3] That duty extends to managing the health risks arising from exposure to asbestos at work.

---

2 The Work Health and Safety Regulations refers to the Regulations in the Commonwealth, NSW, Queensland, SA, Tasmania, ACT and NT based on the Model Work Health and Safety Regulations.

3 Section 19. Unless otherwise specified, section and clause references are to the *Model Work Health and Safety Act* and *Work Health and Safety Regulations* respectively which have been adopted by the Commonwealth and all Australian states and territories save for Victoria and Western Australia.

Workers include employees, contractors, employees of contractors, subcontractors, employees of subcontractors, labour hire workers, apprentices, trainees, students on work experience, volunteers and anyone else who performs work.[4]

In addition, a PCBU with management or control of a workplace has a duty to ensure, so far as reasonably practicable, that the workplace is safe and free from risks.[5] That duty clearly extends to managing the risks arising from any asbestos at that workplace. The landlord, property manager and PCBU occupying commercial premises are all relevant PCBUs with management or control of those premises. All have concurrent and overlapping obligations in relation to managing any risk arising from the presence of asbestos at those premises.

The duty to ensure health and safety, requires the duty holder to eliminate risks to health and safety. If it is not reasonably practicable to eliminate the risks, the risks must be minimised so far as reasonably practicable. This is important in the context of asbestos management. While the WHS Regulations do not require the eradication of asbestos in the workplace, the effect of the duty of care is that asbestos must be removed if it is reasonably practicable to do so.

¶2-030 Reasonable practicability

The duty of the PCBU is limited by reasonable practicability. Reasonable practicability is a balancing exercise. It means doing what is reasonably able to be done in relation to ensuring health and safety, taking into account and weighing up all relevant matters, including:

- the likelihood of the hazard or the risk concerned occurring
- the degree of harm that might result from the hazard or the risk
- what the person concerned knows, or ought reasonably to know, about the hazard or the risk
- ways of eliminating or minimising the risk
- the availability and suitability of ways to eliminate or minimise the risk, and
- as a last step, the cost associated with available ways of eliminating or minimising the risk, including whether it is grossly disproportionate to the risk.[6]

The concept of reasonable practicability is not new in the law. It has been around for over a century and there is a large body of law interpreting that weighing-up exercise. It is a point-in-time assessment. That is, the weighing-up exercise is considered from the perspective of what was known about the risk (in terms of both likelihood and consequences) and the relevant controls and costs at the point in time the relevant decision was made which gave rise to the alleged breach.

4 Section 7.

5 Section 20.

6 Section 17.

The challenge in relation to the application of this principle to the management of asbestos risks is that a great deal is known about the health effects of asbestos exposure including the devastating terminal illnesses associated with such exposure. On the other hand, the controls are equally well known. In particular, elimination, through asbestos removal, is a known and obvious available control. It is remarkable therefore that so many risk assessments do not come to the conclusion that elimination is a reasonably practicable step. It is likely that many such assessments place too much emphasis on the cost of elimination and insufficient emphasis on the risk side of the weighing-up exercise. In that respect, they are likely to be in error and exposing the relevant duty holder to breaches of their duty of care. In many situations, not only is elimination reasonably practicable, it is often the most cost-effective in the long term given the additional administrative costs of maintaining asbestos registers, the additional costs associated with ongoing maintenance activities at the premises given the continuing presence of asbestos and the inevitability of the need to remove the asbestos as its condition deteriorates over time.

¶2-040 Exposure to airborne asbestos at workplace

Since the risk to health of persons arises from exposure to airborne asbestos fibres, a PCBU must ensure that exposure of a person at the workplace to airborne asbestos is eliminated, so far as is reasonably practicable.

If it is not reasonably practicable to eliminate exposure to airborne asbestos, exposure must be minimised so far as is reasonably practicable.[7] In that regard, a PCBU must ensure that the exposure standard for asbestos is not exceeded at the workplace.[8]

¶2-050 Prohibition on work involving asbestos or ACM

The starting point in relation to asbestos is that there is a prohibition against work involving asbestos unless the worker is licensed to do so. A PCBU must not carry out, or direct or allow a worker to carry out, work involving asbestos.[9] Work "involves" asbestos if the work involves manufacturing, supplying, transporting, storing, removing, using, installing, handling, treating, disposing of or disturbing asbestos or ACM.[10] Furthermore, a PCBU that commissions the removal of asbestos must ensure

7 Clause 420.

8 Clause 420(2). There is an exception in relation to an asbestos removal area: (a) that is enclosed to prevent the release of respirable asbestos fibres in accordance with cl 477, and (b) in which negative pressure is used in accordance with that clause.

9 Clause 419.

10 Clause 419(2). The following are exceptions: (a) genuine research and analysis; (b) sampling and identification in accordance with the Regulations; (c) maintenance of, or service work on, non-friable asbestos or ACM, fixed or installed before 31 December 2003, in accordance with the Regulations; (d) removal or disposal of asbestos or ACM, including demolition, in accordance with the Regulations; (e) the transport and disposal of asbestos or asbestos waste in accordance with the *Protection of the Environment Operations Act 1997;* (f) demonstrations, education or practical training in relation to asbestos or ACM; (g) display, or preparation or maintenance for display, of an artefact or thing that is, or includes, asbestos or ACM; (h) management in accordance with this Regulation of in situ asbestos that was installed or fixed before 31 December 2003; (i) work that disturbs asbestos during

that the asbestos removal work is carried out by a licensed asbestos removalist who is licensed to carry out the work.[11] That means if there is asbestos at premises, the PCBU can't simply engage any contractor to perform maintenance work. The contractor must be licensed to perform asbestos removal work. Indeed, the work must be performed in a manner and using equipment that minimises the disturbance of asbestos fibres. This is why in the long term, elimination is often a far more cost-effective solution since it reduces the ongoing maintenance costs of the premises as well as eliminates the asbestos risks.

¶2-060 Reporting

Exposure to airborne asbestos fibres exceeding the exposure standards is a "notifiable incident".[12]

Asbestos incidents are notifiable incidents if they fall in one of the following three categories: (a) they cause the death of a person, (b) cause a "serious injury or illness" to a person or (c) are a "dangerous incident". Because of the latent effect of asbestos illnesses, more often than not, the notification trigger of an asbestos incident is the third category. However, it is important to note that there is no time limitation on the notification requirement in relation to death or illness. A death of a person is a notifiable incident if the death arose out of the conduct of the business or undertaking regardless of where or when the death occurs.[13] The obligation is triggered by the PCBU becoming aware of the incident. That is, if a PCBU becomes aware that a person has died from exposure to asbestos arising from the PCBU's business or undertaking, the PCBU must notify the incident to the Regulator. A similar obligation arises in relation to serious illnesses. An incident will be notifiable if it causes a "serious injury or illness" to a person. That is defined to include an injury or illness requiring the person to have medical treatment within 48 hours of exposure to a substance.

An asbestos incident will be "dangerous incident" if the incident exposes a worker or any other person to a serious risk to a person's health or safety emanating from an immediate or imminent exposure to an "uncontrolled escape, spillage or leakage of a substance".[14]

---

mining operations that involve the extraction of, or exploration for, a mineral other than asbestos; and (j) laundering asbestos contaminated clothing in accordance with the Regulations: cl 419(3). Further exceptions apply if the regulator approves the method adopted for managing risk associated with asbestos: cl 419(4); soil that a competent person has determined (i) does not contain any visible ACM or friable asbestos, or (ii) if friable asbestos is visible—does not contain more than trace levels of asbestos determined in accordance with *AS 4964:2004 (Method for the qualitative identification of asbestos in bulk samples)* or (b) naturally occurring asbestos managed in accordance with an asbestos management plan prepared under cl 432: cl 419(5). Note Pt 8.9—Asbestos-Related Work—applies to work exempted by cl 419(3)-(5).

11 Clause 458(1). That obligation does not apply if the asbestos to be removed is: (a) 10 m^2 or less of non-friable asbestos or asbestos contaminated dust or debris (ACD) associated with the removal of that amount of non-friable asbestos, or (b) ACD that is not associated with the removal of friable or non-friable asbestos and is only a minor contamination: cl 458(2). In that case, the asbestos removal work must be undertaken by a competent person: cl 458(3).

12 Section 38.

13 Section 38.

14 Section 37(a).

Site preservation requirements apply automatically to notifiable incidents and are also lifted automatically by the attendance of the inspector on site unless a non-disturbance notice is issued by the inspector.[15] Such a notice may apply up to 7 days but the inspector may issue subsequent notices.[16]

These site preservation requirements, however, do not prevent any action to assist an injured person or remove a deceased person, or any action that is essential to make the site safe or to minimise the risk of a further notifiable incident. Nor do the restrictions prevent actions associated with a police investigation or for which an inspector or the regulator has given permission.

If an incident is a notifiable incident, the incident must be reported to the Regulator by the fastest possible means.[17] The notification must be given by telephone or in writing (including email or facsimile).[18] If notification of an incident is done by telephone, the person giving notice must give the details of the incident requested by the Regulator and if required by the Regulator, give a written notice of the incident within 48 hours of that requirement being made. A written notice must be in a form, or contain the details, approved by the Regulator. The PCBU must keep a record of each notifiable incident for at least 5 years from the day that notice of the incident is given to the Regulator.

¶2-070 Duty of officers

Officers of companies[19] have a duty to exercise duty diligence to ensure compliance by their companies with work health and safety duties and obligations.

Due diligence is defined to include taking reasonable steps to:

- **Know** — acquire and keep up-to-date knowledge of work health and safety matters. In the context of asbestos, this includes knowing the extent of the organisation's legal obligations in relation to asbestos management. This element of the duty would extend to ensuring adequate due diligence in relation to any property acquisition or lease acquisitions to ensure that the properties do not contain asbestos.
- **Understand** — gain an understanding of the nature of the operations of the business or undertaking and generally of the hazards and risks associated with those operations. If there is asbestos in any of the organisation's workplaces, risks arising from asbestos exposure must be identified as a critical risk and a clear asbestos management plan developed for the relevant workplace which will articulate the timeframe for eradication as well as short-term critical controls

15 Section 198.

16 Section 201.

17 Section 38(2).

18 Section 38(3).

19 Officers are defined by reference to s 9 of the Corporations Act. See Tooma M (2017) Due Diligence: Duty of Officers, 2nd edn, CCH, pp 21–29.

that will be implemented and maintained to control the risk of exposure to asbestos.

- **Resource** — ensure that the person conducting the business or undertaking has available for use, and uses, appropriate resources and processes to eliminate or minimise health and safety risks. In relation to asbestos, this element requires development of a plan and allocation of adequate resources allocated to eradicate any asbestos associated with the organisation's undertaking. In the short term, it requires ensuring there are adequate processes for systematically identifying asbestos, developing and maintaining an asbestos register, labelling the asbestos, developing and maintaining an asbestos management plan and providing adequate information, instruction and training in relation to the management of asbestos risks in relation to the organisation's undertaking.
- **Monitor** — ensure that the person conducting the business or undertaking has appropriate processes for receiving and considering information regarding incidents, hazards and risks and responding in a timely way to that information. In the context of asbestos, this requires monitoring the effectiveness of the processes identified in the "resources element", ensuring there are adequate processes for responding to an asbestos incident and ensuring that such processes are effectively followed.
- **Comply** — to ensure that the person conducting the business or undertaking has, and implements, processes for complying with any duty or obligation they have under the Work Health and Safety Act. In relation to asbestos, this includes obligations in relation to the asbestos register, asbestos labelling, an asbestos management plan, information instruction and training, and in the event of asbestos removal work, requiring a license ensuring that an appropriately licensed asbestos removalist is engaged to undertake the removal work. There are also relevant obligations in relation to air monitoring, health monitoring and obtaining a relevant clearance certificate for the asbestos removal work which may also apply.
- **Verify** — verify the provision and use of the resources and processes referred to above. Where there is identified asbestos on premises used in the undertaking of the organisation, the verification element would require regular asbestos audits to be conducted to ensure that the risk is effectively managed in the short term until eradication can take place.[20]

Given the health impact of asbestos, it is difficult to see how an officer of an organisation that operates within premises that have ACM can discharge their due diligence duty unless a plan is developed for the elimination of asbestos from its undertaking as soon as reasonably practicable.

20 Section 27(5).

CHAPTER 3

HOW TO PROACTIVELY MANAGE ASBESTOS IN THE WORKPLACE?

Roofing contractor prosecuted for exposing workers and public to asbestos

Roofmasters Pty Ltd was a roofing and waterproofing business. It was engaged to repair a commercial building's roof which sustained storm damage. The building contained asbestos.

At that time the premises was leased to a business which sold and repaired tools. The quote for repairs included an allowance for working with asbestos. This included authorised suits, gloves and masks for workers, plastic fabrication frame to catch debris from asbestos and disposal of all waste at an approved centre. A safe work method statement identifying the asbestos hazard and planned controls had been prepared.

Workers from Roofmasters attended the site on two occasions. Those workers had not been advised of the presence of asbestos. They were also not provided with the safe work method statement. The workers were not trained in identifying asbestos or its safe handling. Powered screw guns were used in the work. No personal protective equipment (PPE) suitable for asbestos related work was used.

Quantities of white dust and other contaminated waste were created by the workers. Some of the work was performed on a weekend and on another occasion when the business was open to the public.

Roofmasters workers, staff from the business leasing the premises and members of the public were exposed and made complaints.

Notices were issued by Workplace Health and Safety, Queensland to clean up the site.

The business leasing the premises ceased operating to the public for over 2 months whilst remediation works were undertaken. Some of their tools could not be properly cleaned and were discarded. The tool store and the owners of the building expended considerable sums of money rectifying the damage.

Acting Magistrate Robert Walker convicted and fined the defendant $100,000 and ordered professional costs of $1,000. In reaching a decision, Magistrate Walker considered the failure of the defendant to implement any controls on site either for its workers or others to be a serious breach of the Act.[1]

The Roofmasters case serves as a useful "what not to do" list when it comes to undertaking work on premises containing asbestos. The site was known to contain asbestos. The work was scoped on that basis. Yet, workers were not informed of the presence of asbestos, were not trained in identifying or handling asbestos and were not issued with appropriate PPE for performing the work. Power tools, prohibited in these circumstances, were used. No precautions were taken to contain the asbestos dust arising from the work. As a result, Magistrate Walker found that Roofmasters' workers, the workers of the leasing contractor and members of the public were exposed to a serious risk to their health and safety.

¶3-010 Asbestos to be identified or assumed at workplace

The starting point of the risk management process is identification of the hazard. A person with management or control of a workplace must ensure, so far as is reasonably practicable, that all asbestos or asbestos containing material (ACM) at the workplace is identified by a "competent person".[2]

Competent person means a person who has acquired through training or experience the knowledge and skills to carry out the relevant task. In relation to providing clearance certificates,[3] that must include not only training or experience in relation to asbestos removal industry practice but also qualifications in the form, or a certificate in relation to the specified Vocational Education Training course for asbestos assessor work, or a tertiary qualification in occupational health and safety, occupational hygiene, science, building, construction or environmental health.[4]

In practice, it is prudent to engage an industrial hygienist to undertake a survey of the premises to identify relevant asbestos and to assist in the development of the asbestos register and asbestos management plan discussed below. The table below, extracted from the *Code of Practice: How to Manage or Control Asbestos in the Workplace*, Safe Work

1 *Work Health and Safety Queensland v Roofmasters Pty Ltd*, Brisbane Magistrate's Court, 20 June 2017, Acting Magistrate Walker.

2 Clause 422(1). Except if the person: (a) assumes that asbestos or ACM is present or (b) has reasonable grounds to believe that asbestos or ACM is not present: Cl 422(3).

3 That is, clearance certificates in relation to asbestos removal under cl 473 discussed below.

4 Clause 5.

Australia, October 2018 (**Code of Practice on Asbestos Management**), provides useful guidance on how to identify asbestos in the workplace.

If material at the workplace cannot be identified but a competent person reasonably believes that the material is asbestos or ACM — the PCBU with control or management of workplace must assume that the material is asbestos.[5]

If part of the workplace is inaccessible to workers and is likely to contain asbestos or ACM — the PCBU with management or control of the workplace must assume that asbestos is present in that part of the workplace.[6] The Code of Practice on Asbestos Management provides the following examples of inaccessible areas that may contain asbestos or ACM:

- a cavity in a building that is completely (or almost completely) enclosed and suspected of containing asbestos (based on where asbestos is located elsewhere in the building) where access is only possible through destruction of part of the walls of the cavity
- the inner lining of an old boiler pressure vessel (information on this type of vessel suggests it contains asbestos) which is not accessible due to the design and operation of the boiler and access can only be via partial destruction of the outer layer
- vinyl tiles that may contain asbestos, which have had a number of layers of non-ACM placed over them and secured, where the layers above have been well secured and require some form of destruction in order to access the vinyl that may contain asbestos
- enclosed riser shafts in multistorey buildings containing cables that may be insulated with ACM
- air-conditioning ducts that may contain asbestos gaskets and linings.[7]

If asbestos or ACM is assumed to be present at a workplace, it is taken to be identified at the workplace.[8]

¶3-020 Analysis of sample

A PCBU with management or control of a workplace may identify asbestos or ACM by arranging for a sample of material at the workplace to be analysed for the presence of asbestos or ACM.[9] If a person with management or control of a workplace arranges for an analysis, the person must ensure that the sample is analysed only by a NATA-accredited laboratory accredited for the relevant test method or a laboratory approved or operated by the Regulator.[10]

5 Clause 422(2)(a).
6 Clause 422(2)(b).
7 Code of Practice, p 17.
8 Clause 422(4).
9 Clause 423(1).
10 Clause 423(2).

Table 3.1: Factors to consider when identifying asbestos (Source: Safe Work Australia (2018) *Code of Practice: How to Manage or Control Asbestos in the Workplace*)

	Factors to consider when identifying asbestos
When was the building constructed?	Asbestos was widely used as construction and insulation material in buildings until the late 1980s when bans on its manufacture and use were put in place. However, the use of asbestos was only completely prohibited on 31 December 2003. As the bans were not absolute prior to 2003 and building materials may have been stockpiled, stored, or recycled and used, it is possible that asbestos may be present in buildings that were constructed before 31 December 2003 and possibly later.
Were there any refurbishments or additions to the building prior to 31 December 2003?	Any refurbishment or extensions to the original building prior to 1990 and potentially before 31 December 2003 may have involved the use of asbestos. Even if the original parts of the building did not contain asbestos, it should not be assumed that subsequent additions have no asbestos. Some building products supplied to Australian workplaces are manufactured in countries where asbestos is still used. While the importation of products containing asbestos was prohibited on 31 December 2003, the importer or supplier should be contacted to confirm if products have been tested to ensure they do not contain asbestos and a test certificate should be obtained.
What type of material was used to construct the building?	The main construction materials used are made from timber, brick, steel and cement sheet. If cement sheet is present and was installed up until 1990, it is likely to contain asbestos bonded to the cement particles. For example, a roof made from corrugated cement sheeting is likely to contain asbestos. Areas of buildings that are prone to wet conditions may contain asbestos in the walls and floors due to its hardiness and waterproofing qualities compared to other materials. For example, bathrooms, toilets and laundries may have asbestos sheeting or vinyl tiles. Likewise, pipes throughout the building that carry water and sewage may also contain asbestos.
Talk to designers, manufacturers or suppliers of plant, or refer to design plans	Asbestos may be present in specific parts of plant, like gaskets, insulation and friction brake products, installed in workplaces before 31 December 2003. However, time limited exemptions were allowed where asbestos was used for specific applications, including: – gaskets for use with saturated steam, superheated steam, or with substances classified as dangerous goods and as being toxic or very toxic (31 December 2004) – gaskets for use in liquid chlorine plant designed to operate at 45°C and 1500 kPa pressure (31 December 2006) – vanes for rotary vacuum pumps or compressors (31 December 2007) – split face seals used to prevent leaks from cooling water pumps in coal-fired power stations (31 December 2007) – diaphragms used in electrolytic cells in existing NaCl electrolysis plants, also known as chlor-alkali plants (31 December 2006), and – mission-critical defence components (31 December 2010).

	Factors to consider when identifying asbestos
	If there is plant in the workplace that was designed, built and installed before 31 December 2003 the supplier, manufacturer or designer of the plant should be consulted to find out if asbestos is present and, if possible, obtain this advice in writing. If this is not possible, review the design plans and seek advice from an experienced engineer or plant designer. Some components, including friction materials, supplied to Australian workplaces are manufactured in countries where asbestos is still used. While importation of products containing asbestos was prohibited on 31 December 2003, the importer or supplier should be contacted to confirm if products have been tested to ensure they do not contain asbestos and a test certificate should be obtained. Quality assurance systems or checks should be in place to confirm whether asbestos is present.
Talk to workers who have worked at the workplace for a long time	Speaking with experienced workers will assist in the identification process as they may be aware of the history of the building, including its age, construction, renovation or repairs, and may know where asbestos is located in the workplace.
Visually inspect the workplace to identify asbestos, ACM and inaccessible areas	A thorough inspection of all areas of the workplace must be conducted, including all buildings, structures, ceiling spaces, cellars, shafts, storage areas and wall cavities. Material needs to be considered to contain asbestos unless proven otherwise if: – it cannot be identified – there is uncertainty as to whether it contains asbestos, and – it is inaccessible. The design plans for a building, structure, ship or plant may assist in identifying inaccessible areas, as would discussion with builders, architects, manufacturers of plant and maintenance workers. Knowledge of materials used in the construction of the building or experience and findings from inspections of similar sections of the building (or similar buildings) may also assist.
Take notes and photographs	Taking notes and photographs while the inspection is being conducted can assist in producing the asbestos register.

¶3-030 Presence and location of asbestos to be indicated

A PCBU with management or control of a workplace must ensure that the presence and location of asbestos or ACM identified at the workplace is clearly indicated.[11] If it is reasonably practicable to do so, the PCBU must indicate the presence and location of the asbestos or ACM by a label.[12] The PCBU should seek the advice of a competent person in relation to the number and locations of labels and warning signs.

11 Clause 424(a).

12 Clause 424(b).

If labels are used to identify the presence and location of asbestos or ACM, the location of labels should be consistent with the location listed in the asbestos register.[13] If a risk assessment suggests asbestos may be disturbed or people are likely to be exposed and it is not reasonably practicable to label asbestos directly, a prominent warning sign must be posted in its immediate vicinity.[14] An example of this may be, if floor tiles have been identified as containing asbestos. If so, an appropriate warning sign may be displayed on an adjacent wall.[15] All warning signs must comply with AS 1319–1994: *Safety signs for the occupational environment*. The Code of Practice on Asbestos Management provides that any areas of a workplace that contain asbestos, including plant, equipment and components, must be signposted with warning signs to ensure the asbestos is not unknowingly disturbed without the correct precautions being taken.[16] These signs must be weatherproof, constructed of lightweight material and adequately secured.[17] Signs must also be placed at all the main entrances to the work areas where asbestos is present.[18] Where it is not reasonably practicable to use labels or warning signs to indicate the presence and location of asbestos or ACM, a permit-to-work system may be used to ensure that contractors are made aware of the presence of asbestos before commencing work.[19]

¶3-040 Asbestos register

A PCBU with management or control of a workplace must ensure that an asbestos register is prepared and kept at the workplace.[20] The asbestos register must be regularly maintained to ensure the information in the register is up to date.[21]

The asbestos register must record any asbestos or ACM identified at the workplace including the date on which the asbestos or ACM was identified, and the location, type and condition of the asbestos or ACM. It must state that "no asbestos or ACM is identified at the workplace" if the person knows that no asbestos or ACM is identified, or is likely to be present from time to time, at the workplace.[22]

A PCBU is not required to prepare an asbestos register for a workplace if a register had already been prepared for that workplace.[23]

A PCBU with management or control of a workplace where an asbestos register is kept must ensure that the register is reviewed and as necessary revised if the asbestos

---

13 Code of Practice, p 19.

14 Ibid.

15 Ibid.

16 Ibid, p 20.

17 Ibid.

18 Ibid.

19 Ibid.

20 Clause 425(1).

21 Clause 425(2).

22 Clause 425(3).

23 Clause 425(4). This requirement does not apply to a workplace if: (a) the workplace is a building that was constructed after 31 December 2003, (b) no asbestos has been identified at the workplace and (c) no asbestos is likely to be present at the workplace from time to time: cl 425(6).

Figure 3.1: Asbestos Labels and Warning signs (Source: Safe Work Australia (2018) *Code of Practice on Asbestos Management*, p 57)

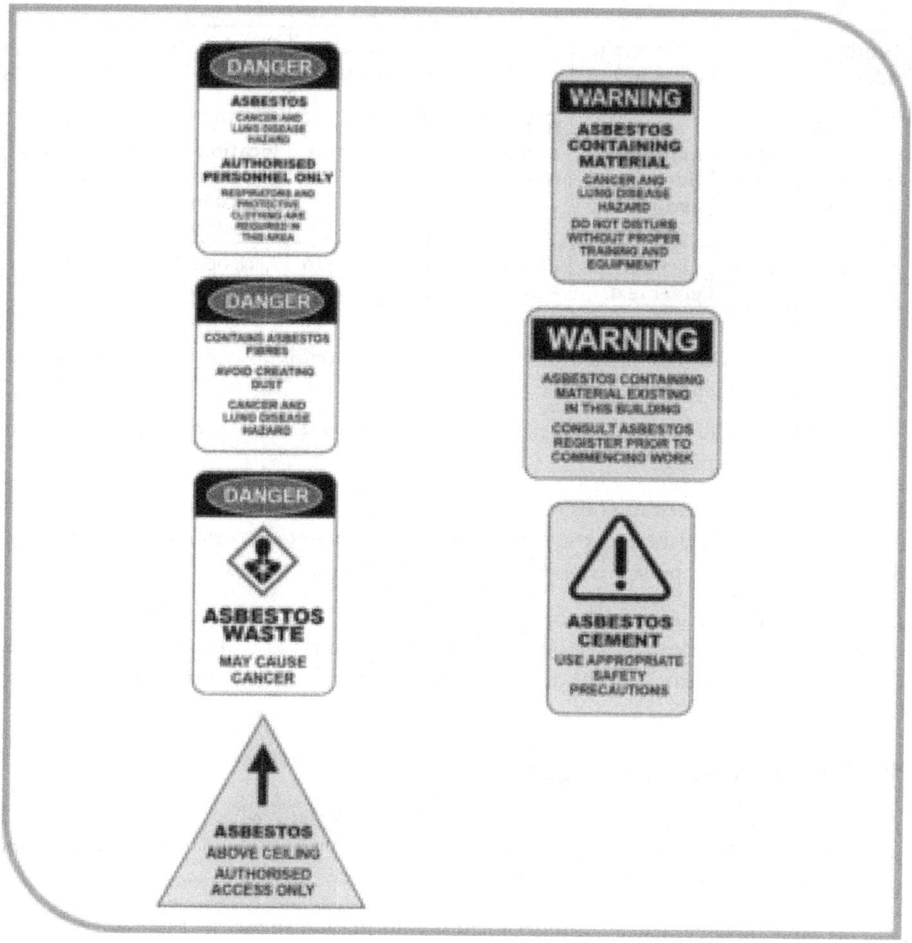

management plan is revised; further asbestos or ACM is identified; or asbestos is removed, disturbed, sealed or enclosed at the workplace.[24]

A PCBU with management or control of a workplace where an asbestos register is kept must ensure that the asbestos register is readily accessible to:

- a worker who has carried out, carries out or intends to carry out, work at the workplace;
- a health and safety representative at the workplace

24 Clause 426.

Table 3.2: Template of asbestos register (Source: Safe Work Australia (2018) *Code of Practice Asbestos Management*, p 59)

Asbestos register					
Workplace address:	Click here to enter text.		**Name of competent person:**	Click here to enter text.	
Date asbestos was identified or assumed to be in the workplace	**Type of material (identified/ assumed)**	**Is it friable or non-friable?**	**Condition of asbestos**	**Specific location of identified or assumed asbestos**	**Is this an inaccessible area?**
Click here to enter text.	Click here to enter text.	Click here to enter text.	Click here to enter text.	Click here to enter text.	Click here to enter text.

- a PCBU who has carried out, carries out or intends to carry out, work at the workplace, and
- a PCBU who has required, requires or intends to require work to be carried out at the workplace.[25]

If a PCBU carries out, or intends to carry out, work at a workplace that involves a risk of exposure to airborne asbestos, the person with management or control of the workplace must ensure that the person is given a copy of the asbestos register.[26]

If a person with management or control of a workplace plans to relinquish management or control of the workplace, the person must ensure, so far as is reasonably practicable, that the asbestos register is given to the person, if any, assuming management or control of the workplace.[27]

¶3-050 Asbestos management plan

A PCBU with management or control of the workplace where asbestos has been identified or is likely to be present must ensure that a written plan for the management of that asbestos is prepared and maintained.[28]

The asbestos management plan must include information about the following:

- the identification of asbestos or ACM
- decisions, and reasons for decisions, about the management of asbestos at the workplace
- procedures for detailing incidents or emergencies involving asbestos or ACM at the workplace,
- workers carrying out work involving asbestos.[29]

25 Clause 427(1).

26 Clause 427(2).

27 Clause 428.

28 Clause 429(2)–(3).

29 Clause 429(4).

A PCBU with management or control of a workplace must ensure that a copy of the asbestos management plan for the workplace is readily accessible to:

- a worker who has carried out, carries out or intends to carry out, work at the workplace;
- a health and safety representative at the workplace;
- a PCBU who has carried out, carries out or intends to carry out, work at the workplace, and
- a PCBU who has required, requires, or intends to require work to be carried out at the workplace.[30]

A PCBU with management or control of a workplace that has an asbestos management plan must ensure that the plan is reviewed and as necessary revised in the following circumstances:

- there is a review of the asbestos register or a control measure;
- asbestos is removed from, or disturbed, sealed or enclosed at, the workplace;
- the plan is no longer adequate for managing asbestos or ACM at the workplace;
- a health and safety representative requests a review;
- at least once every 5 years.[31]

A health and safety representative for workers at a workplace may request a review of an asbestos management plan if the representative reasonably believes that one of the above circumstances may affect the health and safety of a member of the work group represented by them and the PCBU has not adequately reviewed the asbestos management plan in response to the circumstance.[32]

Failure to have asbestos registers results in fine

On 1 March 2014, a person was observed undertaking refurbishment work on their property which included removing soffit sheeting from the underside of an awning at a commercial property. The soffit sheeting contained asbestos. Dust and debris were seen to fall onto the footpath. He was observed removing batons and pieces of broken sheeting. He was also observed sweeping up dust and debris with a broom. A group of shops were at this location, some with tenants. Pedestrians were walking under the soffits while the work was being done.

A witness, recognising the risk associated with these activities, contacted the Regulator.

The person advised the Regulator that he owned the property and did the work on the assumption that the soffits may contain asbestos. He advised that two panel sheets had fallen from the shops onto the pavement due to white ant damage. He advised that he

30 Clause 429(5).

31 Clause 430(1).

32 Clause 430(2).

picked up pieces of material and swept up the area and put the debris into plastic bags in his trailer. He indicated that he would call the Council to arrange disposal. The inspector provided advice to him, particularly that he needed to have a license if he intended to remove more than 10 m^2 of ACM.

Three days later, another inspector attended the incident site. He observed that the awning soffit sheeting of the shops had been recently altered and replaced with what appeared to be low-profile ribbed metal sheeting. He measured the soffit removal area (total of 62.79 m^2) and seized samples.

He was charged with breaching the *Work Health and Safety Regulation 2011* (Qld). In particular, he was charged with failing to ensure an asbestos register was prepared and kept at the workplace (cl 425(1)) and failing to ensure a written asbestos management plan for the workplace was prepared (cl 492(2)). He was also charged for carrying out work without being authorised by not being the holder of an asbestos removal license (s 43(1) of the Work Health and Safety Act by virtue of cl 487(1)).

He pleaded guilty in the Bowen Magistrates Court. Magistrate Simon Young fined the defendant $750. No conviction was recorded.[33]

¶3-060 Naturally occurring asbestos

A PCBU with management or control of a workplace must manage risks to health and safety associated with naturally occurring asbestos at the workplace.[34] Naturally occurring asbestos is generally found when building roads, working on construction sites and undertaking excavation activities.

If naturally occurring asbestos is identified at a workplace or is likely to be present at a workplace, a PCBU with management or control of the workplace must ensure that an asbestos management plan for the workplace is prepared in relation to the naturally occurring asbestos.[35] The asbestos management plan must be maintained to ensure the information in the plan is up to date.[36] The plan must include information about the following:

- the identification of naturally occurring asbestos;
- decisions, and reasons for decisions, about the management of naturally occurring asbestos at the workplace; procedures for detailing incidents or emergencies involving naturally occurring asbestos at the workplace; and
- workers carrying out work involving naturally occurring asbestos.[37]

A copy of the asbestos management plan for naturally occurring asbestos at the workplace must be readily accessible to affected workers, health and safety representatives of those

33 *Work Health and Safety Queensland v E195327*, Bowen Magistrates Court, 15 March 2016, Magistrate Young.

34 Clause 431.

35 Clause 432(2).

36 Clause 432(3).

37 Clause 432(4).

workers, other PCBUs carrying out work at the workplace or intending to do so.[38] The asbestos management plan for naturally occurring asbestos must be reviewed and as necessary revised if the plan is no longer adequate for managing naturally occurring asbestos at the workplace.[39]

The PCBU must provide training in the hazards and risks associated with naturally occurring asbestos for workers who carry out work where naturally occurring asbestos is likely to be found.[40]

¶3-070 Asbestos training

A PCBU must ensure that workers engaged by them, whom the person reasonably believes may be involved in asbestos removal work or in the carrying out of asbestos-related work, are trained in the identification and safe handling of, and suitable control measures for, asbestos and ACM.[41]

A record is kept of the training undertaken by the worker while the worker is carrying out the work and for 5 years after the day the worker ceases working for the person.[42] The record must be available for inspection by a Regulator.[43]

Figure 3.1: Summary of asbestos management obligations of PCBU with management or control of workplace

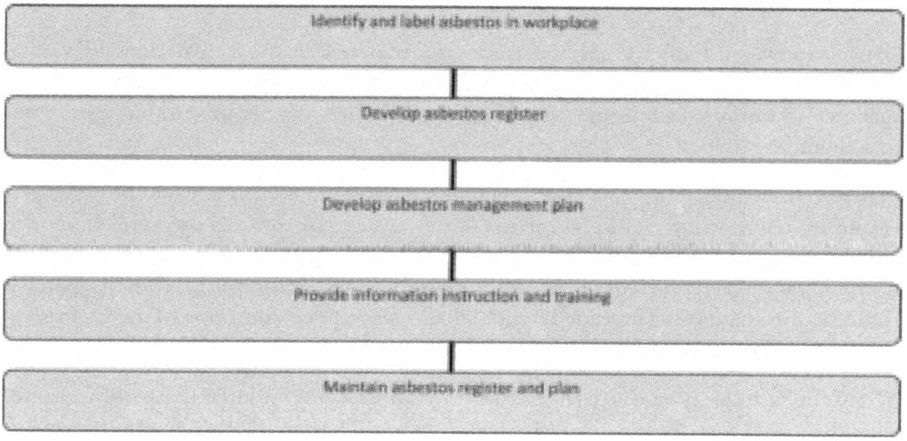

38 Clause 432(5).

39 Clause 433.

40 Clause 434.

41 Clause 445(1). This does not apply to workers involved in licensed asbestos removal work since additional requirements apply to such workers.

42 Clause 445(3).

43 Clause 445(4).

¶3-080 Prohibited use of equipment in relation to asbestos

The use of certain equipment has the potential to disturb asbestos, causing the release of airborne asbestos fibres. As such the PCBU must ensure that such equipment is not used on asbestos or ACM. In that regard the PCBU must not use, direct or allow a worker to use high pressure water spray or compressed air equipment on asbestos or ACM.[44] They must also ensure that certain equipment is controlled in relation to its use. Specifically, the PCBU must not use, direct or allow a worker to use power tools, brooms or any other implements that cause the release of airborne asbestos into the atmosphere on asbestos or ACM unless the use of the equipment is controlled. This control is achieved by:

- enclosing the equipment during its use
- using equipment which is designed to capture or suppress airborne asbestos
- using the equipment in a way that is designed to capture or suppress airborne asbestos safely.[45]

The Code of Practice provides that manually operated (non-powered) hand tools should be used wherever possible.[46] The Code provides that if they are not suitable for the task or will not provide sufficient physical force to perform the required operation, low-speed, battery-powered tools that are able to be used in conjunction with wet methods for dust control are preferred.[47]

The Code further provides that:

"Battery-powered tools should be fitted with a Local Exhaust Ventilation (LEV) dust control hood wherever possible. If an LEV dust control hood cannot be attached then shadow vacuuming techniques should be used. Where this is unsuitable, pastes and gels should be used".[48]

Where power tools with dust suppression/extraction are used, the Code provides that "exposure monitoring should be carried out to ensure the controls used are effective in reducing the generation of fibres. It is good practice to ensure that the levels of airborne fibres do not exceed one half of the exposure standard (0.1 fibres/mL). If more than half the exposure standard is exceeded, work should be stopped and improvements made to the controls being used".[49]

The Code further provides that HEPA filtered H-Class industrial vacuum cleaners should be used for asbestos-related work and should comply with the requirements in AS/NZS 60335.2.69:2017: *Household and similar electrical appliances—Safety: Particular requirements for wet and dry vacuum cleaners, including power brush, for commercial use* (IEC

44 Clause 446(1). An exception applies to fire-fighting purposes: cl 446(2).

45 Clause 446(3).

46 Code of Practice, p 46.

47 Ibid.

48 Ibid.

49 Ibid.

60335–2–69 Ed 5 MOD).[50] It provides that they should also undergo annual filtration efficiency testing.[51]

Household vacuum cleaners must never be used where asbestos is or may be present, even if they have a HEPA filter.[52]

Use of power tools results in prosecution

A building subcontractor was contracted by a painting company to clean the roof of residential premises prior to painting. There were no conversations between the subcontractor and the contractor regarding the composition of the roof, safe procedures or equipment to use.

On 15 September 2016, the subcontractor accessed the roof and sprayed it with a hydrochloride solution and used a "whirl-a-way" surface cleaner attached to a high-pressure washer to clean the back section of the roof. This resulted in three adjacent residential properties being contaminated with debris.

The subcontractor was prosecuted for used high-pressure water to spray ACM in contravention of s 446(1)(a) of the Work Health Safety Regulation 2011.

It pleaded guilty in the Townsville Magistrates Court. Magistrate Scott Luxton fined the defendant $1,000 plus costs with a 1-year good behaviour bond of $1,000. No conviction was recorded.[53]

¶3-090 Demolition and refurbishment

The PCBU with management or control of a workplace must ensure that, before demolition or refurbishment of that workplace is carried out, the asbestos register for the workplace is reviewed.[54] If the register is inadequate having regard to the proposed demolition or refurbishment, the register must be revised.[55] The PCBU must ensure that the worker who carries out the demolition or refurbishment is given a copy of the asbestos register before the demolition or refurbishment is commenced.[56] Conversely, a PCBU who carries out demolition or refurbishment at a workplace must obtain a copy of the asbestos register from the person with management or control of the workplace, before the person commences the demolition or refurbishment.[57]

If there is no asbestos register, the PCBU who is to carry out the demolition or refurbishment must not carry out the demolition or refurbishment until the structure

50 Ibid.

51 Ibid at 47.

52 Ibid.

53 *Work Health and Safety Queensland v E233396*, Townsville Magistrates Court, 15 January 2018, Magistrate Luxton.

54 Clause 448(a).

55 Clause 448(b).

56 Clause 449.

57 Clause 450.

Figure 3.2: Safe Work Practices — Drilling ACM (Source: Safe Work Australia (2018) *Code of Practice: How to Manage Asbestos in the Workplace*, pp 63–66)

	Safe work practice 1 — Drilling of ACM
	The drilling of asbestos cement sheeting or low density asbestos fibre board can release asbestos fibres into the atmosphere, so precautions must be taken to protect the drill operator and other persons from exposure to these fibres. A hand drill is preferred to a battery-powered drill, because the quantity of fibres is drastically reduced if a hand drill is used.
Equipment that may be required prior to starting work (in addition to what is needed for the task)	– A non-powered hand drill or a low-speed battery-powered drill or drilling equipment. Battery-powered drills should be fitted with a local exhaust ventilation (LEV) dust control hood wherever possible. If an LEV dust control hood cannot be attached then shadow vacuuming techniques should be used. If this is not possible, other dust control methods such as pastes and gels should be used. – Disposable cleaning rags – A bucket of water, or more as appropriate, and/or a misting spray bottle – Adhesive (cloth or duct) tape – Sealant – Spare PPE – A thickened substance such as wallpaper paste, shaving cream or hair gel – Heavy duty plastic (200 μm polyethylene) sheeting – A suitable asbestos waste container (eg heavy duty plastic bags or a drum, bin or skip lined with heavy duty plastic sheeting) – Warning signs and/or PVC barrier tape – A HEPA filtered H-Class industrial vacuum cleaner – A sturdy paper, foam or thin metal cup, or similar (for work on overhead surfaces only)
PPE	– Protective clothing and RPE (see AS/NZS 1715:2009: *Selection, use and maintenance of respiratory protective equipment* and AS/NZS 1716:2012: *Respiratory protective devices*). It is likely that a class P1 or P2 half-face respirator will be adequate for this task, provided the recommended safe work procedure is followed.
Preparing the asbestos work area	– If the work is to be carried out at height, appropriate precautions must be taken to prevent falls. – Ensure appropriately labelled heavy duty plastic asbestos waste disposal bags are available. – Carry out the work with as few people present as possible. – Segregate the asbestos work area to ensure unauthorised personnel are restricted from entry (eg close door and/or use warning signs and/or barrier tape at all entry points). The distance for segregation should be determined by a risk assessment. – If drilling a roof from outside, segregate the area below. – If access is available to the rear of the asbestos cement, segregate this area as well as above. – If possible, use heavy-duty plastic sheeting, secured with adhesive (cloth or duct) tape, to cover any surface within the asbestos work area that could become contaminated.

	Safe work practice 1 — Drilling of ACM
	– Ensure there is adequate lighting. – Avoid working in windy environments where asbestos fibres can be redistributed. – If using a bucket of water, do not resoak used rags in the bucket, as this will contaminate the water. Instead, either fold the rag so a clean surface is exposed or use another rag.
Drilling vertical surfaces	– Tape both the point to be drilled and the exit point, if accessible, with a strong adhesive (cloth or duct) tape to prevent the edges crumbling. – If possible, use local exhaust ventilation or shadow vacuuming techniques to capture the small amounts of debris created during drilling. If this is not possible, cover the drill entry and exit points (if accessible) on the asbestos with a generous amount of thickened substance. – Drill through the tape. – Use damp rags to clean off paste (if used) and debris from the wall and drill bit. – Dispose of the rags as asbestos waste as they will contain asbestos dust and fibres. – Seal the cut edges with sealant. – If a cable is to be passed through, insert a sleeve to protect the inner edge of the hole.
Drilling overhead horizontal surfaces	– Mark the point to be drilled. – Drill a hole through the bottom of the cup. – Fill or line the inside of the cup with shaving cream, gel or a similar thickened substance. – Put the drill bit through the hole in the cup so that the cup encloses the drill bit, and make sure the drill bit extends beyond the lip of the cup. – Align the drill bit with the marked point. – Ensure the cup is firmly held against the surface to be drilled. – Drill through the surface. – Remove the drill bit from the cup, ensuring that the cup remains firmly against the surface. – Remove the cup from the surface. – Use damp rags to clean off the paste and debris from the drill bit. – Dispose of the rags as asbestos waste, as they will contain asbestos dust and fibres. – Seal the cut edges with sealant. – If a cable is to be passed through, insert a sleeve to protect the inner edge of the hole.
Decontaminating the asbestos work area and equipment	– Use damp rags to clean the equipment. – Carefully roll or fold any plastic sheeting used to cover any surface within the asbestos work area, so as not to spill any dust or debris that has been collected. – If necessary, use damp rags and/or a HEPA filtered H-Class industrial vacuum cleaner to clean any remaining visibly contaminated sections of the asbestos work area.

	Safe work practice 1 — Drilling of ACM
	– Place debris, used rags, plastic sheeting and other waste in appropriately labelled heavy duty plastic asbestos waste disposal bags or wrap it in a double layer of heavy duty plastic which is then sealed using adhesive (cloth or duct) tape and appropriately labelled as asbestos waste. – Wet wipe the external surfaces of the asbestos waste disposal bags or wrapping to remove any adhering dust before they are removed from the asbestos work area.
Personal decontamination should be carried out in a designated area	– If disposable coveralls are worn, clean the coveralls while still wearing RPE using a HEPA filtered H-Class industrial vacuum cleaner, damp rag or fine-water spray. RPE can be cleaned with a wet rag or cloth. – While still wearing RPE, remove coveralls, turning them inside-out to entrap any remaining contamination and then place them into appropriately labelled asbestos waste disposal bags or wrap them in a double layer of heavy duty plastic which is then sealed using adhesive (cloth or duct) tape and appropriately labelled as asbestos waste. – Remove RPE. If non-disposable, inspect it to ensure it is free from contamination, clean it with a wet rag and store in a clean container. If disposable, cleaning is not required but RPE should be placed in an appropriately labelled heavy duty plastic asbestos waste disposal bag or wrap it in a double layer of heavy duty plastic which is then sealed using adhesive (cloth or duct) tape and appropriately labelled as asbestos waste. Refer to the Code of Practice: *How to safely remove asbestos* for more information.
Clearance procedure	– Visually inspect the asbestos work area to make sure it has been properly cleaned. – Clearance air monitoring is not normally required for this task. – Ensure asbestos waste is transported and disposed of in accordance with the relevant state or territory Environment Protection Authority (EPA) requirements at a site licensed by the EPA. Refer to the Code of Practice: *How to safely remove asbestos* for more information.

or plant has been inspected to determine whether asbestos or ACM is fixed to or installed in the structure or plant.[58] This determination is undertaken by a competent person.[59] If the competent person is uncertain whether or not asbestos is present or the relevant part of the structure or plant is inaccessible, the PCBU must assume that it is.[60] If asbestos or ACM is determined or assumed to be fixed to or installed in the structure or plant, the PCBU must inform the occupier of the premises and owner of the premises

58 Clause 451(2).

59 Clause 451(3).

60 Clause 451(4).

(if the place is residential) or the person with management or control of the workplace (if a workplace).[61]

The PCBU with management or control of the workplace, or of the structure or plant, must ensure that all asbestos that is likely to be disturbed by the demolition is identified and so far as is reasonably practicable, removed before the demolition is commenced.[62] If the premises are residential premises, that obligation falls on the PCBU carrying out the work.[63] A similar obligation applies in relation to refurbishments.[64] As with demolition, if the premises are residential, the obligation falls on the PCBU undertaking the work.[65]

In case of an emergency[66] where a structure or plant at the workplace must be demolished, the PCBU with management or control of the workplace must ensure, so far as is reasonably practicable, that before the demolition is commenced, a procedure is developed that will, so far as is reasonably practicable, reduce the risk of exposure of workers and persons in the vicinity of the demolition site to asbestos to below the exposure standard.[67] The asbestos register for the workplace must be considered in the development of this procedure.[68] The regulator must be given written notice about the emergency immediately after the person becomes aware of the emergency and before the demolition is commenced.[69] If the premises is residential these obligations fall on the PCBU undertaking the work.[70]

61 Clause 451.

62 Clause 452(3). Excludes emergencies and residential properties: cl 452(2). The obligation also does not apply if the purpose of the demolition is to gain access to the asbestos: cl 452(4).

63 Clause 453. Excludes emergencies: cl453(2). The obligation also does not apply if the purpose of the demolition is to gain access to the asbestos: cl453(3).

64 Clause 456.

65 Clause 457.

66 Emergency means that the structure or plant is structurally unsound, or collapse of the structure or plant is imminent: cl 454(4).

67 Clause 454(2)(a).

68 Clause 454(2)(b).

69 Clause 454(3).

70 Clause 455.

Figure 3.3: Safe Work Practices — Sealing, painting, coating and cleaning of asbestos cement products (Source: Safe Work Australia (2018) *Code of Practice: How to Manage Asbestos in the Workplace*, pp 67–69)

	Safe work practice – Sealing, painting, coating and cleaning of asbestos cement products
	These tasks should only to be carried out on asbestos cement products that are in good condition. For this reason, the ACM should be thoroughly inspected before starting the work. There is a risk to health if the surface of asbestos cement sheeting is disturbed (eg from hail storms and cyclones) or if it has deteriorated as a result of aggressive environmental factors such as pollution. If it is so weathered that its surface is cracked or broken, the asbestos-cement matrix may be eroded, increasing the likelihood that asbestos fibres will be released. **If treatment is considered essential, a method that does not disturb the matrix should be used. Under no circumstances should asbestos-cement products be water blasted or dry sanded in preparation for painting, coating or sealing.**
Equipment that may be required prior to starting work (in addition to what is needed for the task)	– Disposable cleaning rags – A bucket of water, or more as appropriate, and/or a misting spray bottle – Sealant – Spare PPE – Heavy duty plastic (200 µm polyethylene) sheeting – A suitable asbestos waste container (eg heavy duty plastic bags or a drum, bin or skip lined with heavy duty plastic sheeting) – Warning signs and/or PVC barrier tape.
PPE	– Protective clothing and RPE (see AS/NZS 1715:2009: *Selection, use and maintenance of respiratory protective equipment* and AS/NZS 1716:2012: *Respiratory protective devices*). It is likely that a class P1 or P2 half face respirator will be adequate for this task, provided the recommended safe work procedure is followed. Where paint is to be applied, appropriate respiratory protection to control the paint vapours/mist must also be considered.
Preparing the asbestos work area	– If work is being carried out at height, precautions must be taken to prevent falls. – Before starting, assess the asbestos-cement for damage. – Ensure appropriately marked asbestos waste disposal bags are available. – Carry out the work with as few people present as possible. – Segregate the asbestos work area to ensure unauthorised personnel are restricted from entry (eg close door and/or use warning signs and/or barrier tape at all entry points). The distance for segregation should be determined by a risk assessment. – If working at height, segregate the area below. – If possible, use heavy duty plastic sheeting secured with adhesive (cloth or duct) tape to cover any floor surface within the asbestos work area which could become contaminated. This will help to contain any runoff from wet sanding methods. – Ensure there is adequate lighting. – If using a bucket of water, do not resoak used rags in the bucket, as this will contaminate the water. Instead, either fold the rag so a clean surface is exposed or use another rag.

	Safe work practice – Sealing, painting, coating and cleaning of asbestos cement products
	– Never use high-pressure water cleaning methods. – Never prepare surfaces using dry sanding methods. Where sanding is required, you should consider removing the asbestos and replacing it with a non-asbestos product. – Wet sanding methods may be used to prepare the asbestos, provided precautions are taken to ensure all the runoff is captured and filtered, where possible. – Wipe dusty surfaces with a damp cloth.
Painting and sealing	– When using a spray brush, never use a high-pressure spray to apply the paint. – When using a roller, use it lightly to avoid abrasion or other damage.
Decontaminating the asbestos work area and equipment	– Use damp rags to clean the equipment. – If required, use damp rags and/or a HEPA filtered H-Class industrial vacuum cleaner to clean the asbestos work area. – Place debris, used rags, plastic sheeting and other waste in appropriately labelled heavy duty plastic asbestos waste disposal bags or wrap it in a double layer of heavy duty plastic which is then sealed using adhesive (cloth or duct) tape and appropriately labelled as asbestos waste. – Wet wipe the external surfaces of the asbestos waste bags/container to remove any adhering dust before they are removed from the asbestos work area.
Personal decontamination should be carried out in a designated area	– If disposable coveralls are worn, clean the coveralls while still wearing RPE using a HEPA filtered H-Class industrial vacuum cleaner, damp rag or fine-water spray. RPE can be cleaned with a wet rag or cloth. – While still wearing RPE, remove coveralls, turning them inside-out to entrap any remaining contamination and then place them into appropriately labelled heavy duty plastic asbestos waste disposal bags or wrap them in a double layer of heavy duty plastic which is then sealed using adhesive (cloth or duct) tape and appropriately labelled as asbestos waste. – Remove RPE. If non-disposable, inspect it to ensure it is free from contamination, clean it with a wet rag and store in a clean container. If disposable, cleaning is not required but RPE should be placed in an appropriately labelled heavy duty plastic asbestos waste disposal bag or wrapped in a double layer of heavy duty plastic which is then sealed using adhesive (cloth or duct) tape and appropriately labelled as asbestos waste. Refer to the Code of Practice: *How to safely remove asbestos* for more information.
Clearance procedure	– Visually inspect the asbestos work area to make sure it has been properly cleaned. – Clearance air monitoring is not normally required for this task. – Ensure asbestos waste is transported and disposed of in accordance with the relevant state or territory Environment Protection Authority (EPA) requirements at a site licensed by the EPA Refer to the Code of Practice: *How to safely remove asbestos* for more information.

CHAPTER 4

WHAT TO DO IF THERE IS AN ASBESTOS INCIDENT?

Incidents involving asbestos exposure are highly complex. In addition to being notifiable incidents, a greater emphasis is required on containment, provision of quality expert information to affected persons, managing stakeholders and managing the physical and psychological impact on individuals exposed to the asbestos.

¶4-010 Notify the Regulator

As discussed in Chapter 2, asbestos exposure incidents are notifiable incidents because they represent an uncontrolled release of a substance that presents an imminent risk to the health of a person. Any incident report must be comprehensive but factual. That is, it is important that incident reports avoid emotive language. The report should specify what is known at the time of the report. If source of the contamination is known, it should be stated. However, if, as often is at the case, the source of the contamination is not known, the report should indicate this. Being precise in relation to such matters is important, not only from the point of view of dealings with the Regulator. Any documents created at this stage will likely be discoverable in any subsequent litigation in relation to compensation arising from the contamination.

¶4-020 Engage an industrial hygienist

Investigating the source and extent of an asbestos contamination requires the specialist expertise of a competent person. While many organisations will have competent persons on staff, given the sensitive nature of an asbestos incident, it is prudent to engage an industrial hygienist to investigate and advise on the issue. Any investigation may include air monitoring to identify the extent of any airborne contaminants and whether the relevant exposure standards have been exceeded.

An industrial hygienist can then develop a remediation plan for the contamination including consideration of removing the relevant asbestos containing material (ACM). Such removal would typically require the engagement of a licensed asbestos removal contractor.

¶4-030 Engaging an occupational physician

One of the issues which must be determined in the aftermath of an incident is the extent to which health monitoring of any affected persons is required or desirable. An occupational physician can play a very constructive role in advising in relation to this issue. For the reasons set out below, even if the specific obligation is not triggered under the legislation, health monitoring may play a useful role in providing peace of mind for affected workers exposed to the asbestos contamination.

In any event, an experienced occupational physician can assist greatly in providing measured medical advice in relation to the mechanism of exposure, possible adverse health impact and the probability of adverse health effects from such exposure. The dissemination of timely, considered health advice to affected workers and their families is crucial in minimising the psychological impact of such incidents. Affected workers often feel a great deal of responsibility or "guilt" for potentially exposing their facilities to harm through their work. Common concerns relate to the potential exposure of family members to asbestos fibres through the washing of work clothes. Considered advice on the real risk of such exposure may go a long way to alleviating such concerns.

¶4-040 Engaging a communication expert

Effective communication in relation to asbestos incidents is crucial. Not only is communication required in relation to external stakeholders such as Regulators, unions, customers and members of the public, effective communication is also required in relation to potentially affected workers and their families. Engaging with relevant internal or external communication experts is very useful in the circumstances. The more the information is provided in relation to the extent of the problem, the health effects of the exposure, the remediation strategy and the support available to anyone affected, the better.

¶4-050 Engaging an asbestos removal contractor

Any remediation strategy in relation to a significant asbestos incident will invariably require the engagement of a licensed asbestos removal contractor to remediate the source of the contamination. As discussed in Chapter 5, any such remediation work will require the approval of the Regulator before it is undertaken. Once the contamination has been addressed, a clearance certificate will be provided.

CHAPTER 5

WHAT TO EXPECT IN ASBESTOS REMOVAL PROCESS?

Contractor prosecuted for asbestos removal without a license

Mr Anthony Rosier was hired by the owner of a Mount Barker, South Australia, residential dwelling to remove the asbestos containing material as part of a project to demolish and remove the whole building.[1]

The owner was a previous client of Mr Rosier. Mr Rosier confirmed to the owner the presence of asbestos, as had been suspected.[2] When inspecting the works to be carried out, he claimed to be licensed to remove asbestos and gave the owner his business card.[3] The business card included a licence number.[4] The stated licence number had previously been Mr Rosier's asbestos removal licence.[5] However, it was cancelled in August 2015 by the Regulator, SafeWork SA, and he was disqualified for reapplying for a new licence for

1 *Martyn Campbell v Anthony Rosier* [2019] SAET 227 (15 November 2019), Deputy President Lieschke at [5].

2 Ibid.

3 Ibid.

4 Ibid.

5 Ibid at [6].

3 years.[6] He was also told he was required to undergo specific training before reapplying.[7] The defendant did not undergo the training and has not reapplied for a licence.[8]

The agreed fee for demolition and asbestos removal was $11,000.[9] The current fee to apply for a new licence is $2,161, and if the application is approved a 3-year licence then attracts an additional $3,852 fee.[10]

The Mount Barker demolition works commenced by late September 2017. By 6 November, reports from people working nearby of concerns about asbestos reached the local Council and were then passed on to SafeWork SA.[11] SafeWork SA inspectors attended the site and spoke to Mr Rosier.[12] He readily admitted to removing asbestos containing material without a licence for doing so.[13] Samples of different materials were taken and were later proven to contain asbestos.[14] These included asbestos cement sheets, asbestos vinyl floor tiles and asbestos-backed vinyl floor covering. There was a mixture of white, blue and brown asbestos in the materials.[15] The asbestos containing material was estimated at 94 m^2 in area, being well above the 10 m^2 maximum area that can be removed without a licence.[16]

Mr Rosier's former business name of Eco Asbestos Removal Services with his attached personal contact details was still published on the internet notwithstanding the loss of loss license. His personal mobile phone still answered automatically with a recorded message referring to Eco Asbestos Removal. Only his former business website of "Eco Demolitions" had been effectively shut down or deleted.[17]

Mr Rosier was prosecuted and fined $8,100.

In determining penalty, His Honour, Deputy President Stephen Leischke noted that "the harm was in not being licensed, and not in the manner in which the work was carried out, other than the failure to erect signage that specifically mentioned asbestos removal was underway. There is no formal complaint about the personal protective equipment (PPE) used by the two workers engaged by the defendant, or of the method of containing the asbestos, or of any need for air monitoring during the removal work".[18]

His Honour noted that:

"The signage offence meant adjoining neighbours, people nearby or other people wishing to enter the site were not alerted to the asbestos removal work being undertaken. Despite no evidence of any actual harm from this failure it could have resulted in people inhaling dangerous asbestos laden dust".[19]

6 Ibid.
7 Ibid.
8 Ibid.
9 Ibid at [5].
10 Ibid at [7].
11 Ibid at [8].
12 Ibid at [9].
13 Ibid at [9].
14 Ibid at [9].
15 Ibid at [9].
16 Ibid at [10].
17 Ibid at [13].
18 Ibid at [15].
19 Ibid at [16].

In relation to the licensing requirements, His Honour noted that:
"The failure to be licensed is far more than just not contributing the set fees to the Government, as the defendant suggests. The failure to be licensed against a background of his licence being cancelled for good cause means the defendant was not able to be monitored or audited regarding his asbestos related work. In the defendant's circumstances this failure was highly relevant".[20]

¶5-010 Asbestos removalist must be licensed

A person conducting a business or undertaking (PCBU) that commissions the removal of asbestos must ensure that the asbestos removal work is carried out by a licensed asbestos removalist who is licensed to carry out the work.[21] "Asbestos removalist" is defined in cl 5 of the WHS Regulations to mean a PCBU who carries out asbestos removal work.

There are two classes of asbestos removal licenses: Class A and Class B. Class A asbestos removal license is defined in cl 5 of the Regulation to mean a license that authorises the carrying out of "Class A" and "Class B" asbestos removal work. Class A asbestos removal work relates to the removal of friable asbestos and asbestos contaminated dust or debris (ACD).[22]

Conversely, Class B asbestos removal license is defined to mean a license that authorises "Class B" asbestos removal work. Class B asbestos removal is asbestos removal related to more than 10 m^2 of non-friable asbestos or asbestos containing material (ACM) or ACD associated with the removal of more than 10 m^2 of non-friable asbestos or ACM.[23]

For a Class A asbestos removal work, the nominated asbestos removal supervisor for the asbestos removal work must be present at the asbestos removal area whenever the asbestos removal work is being carried out.[24] For Class B asbestos removal work, the nominated asbestos removal supervisor for the asbestos removal work must be readily available to a worker carrying out asbestos removal work whenever the work is being carried out.[25]

Workers undertaking asbestos removal work must hold a certification in relation to the specified Vocational Education Training (VET) course for asbestos removal relevant

20 Ibid at [17].

21 Clause 458(1). That obligation does not apply if the asbestos to be removed is: (a) 10 m^2 or less of non-friable asbestos or ACD associated with the removal of that amount of non-friable asbestos, or (b) ACD that is not associated with the removal of friable or non-friable asbestos and is only a minor contamination: cl 458(2). In that case, the asbestos removal work must be undertaken by a competent person: cl 458(3).

22 Clause 485. In relation to ACD, cl 486 provides an exception relation to ACD associated with non-friable asbestos and ACD associated with friable asbestos if it is a minor contamination.

23 Clause 487.

24 Clause 459(a).

25 Clause 459(b).

Table 5.1: Licence requirements for asbestos removal work (Source: Safe Work Australia (2018) *Code of Practice: How to Safely Remove Asbestos*)

Type of licence	What asbestos can be removed?
Class A	Can remove any amount or quantity of asbestos or ACM, including: – any amount of friable asbestos or ACM – any amount of ACD – any amount of non-friable asbestos or ACM.
Class B	Can remove: – any amount of non-friable asbestos or ACM – any amount of ACD associated with the removal of non-friable asbestos or ACM.
No licence required	Can remove: – up to 10 m^2 of non-friable asbestos or ACM – ACD that is: ■ associated with the removal of less than 10 m^2 of non-friable asbestos or ACM ■ not associated with the removal of friable or non-friable asbestos and is only a minor contamination.

to the class of licensed asbestos removal work to be carried out by the worker.[26] In addition, a licensed asbestos removalist must provide appropriate training to a worker carrying out licensed asbestos removal work at a workplace to ensure that the work is carried out in accordance with the asbestos removal control plan for the workplace.[27]

A licensed asbestos removalist must keep a record of the training undertaken by a worker carrying out licensed asbestos removal work while the worker is carrying out licensed asbestos removal work and for 5 years after the day the worker stopped carrying out licensed asbestos removal work for the removalist.[28] The training records must be readily accessible at the asbestos removal area and available for inspection by the Regulator.[29]

A licensed asbestos removalist must give information regarding the health risks and health effects associated with exposure to asbestos and details of health monitoring to anyone likely to be engaged to carry out licensed asbestos removal work.[30]

A licensed asbestos removalist must obtain a copy of the asbestos register for a workplace before the removalist carries out asbestos removal work at the workplace.[31]

A licensed asbestos removalist must prepare an asbestos removal control plan for any licensed asbestos removal work the removalist is commissioned to undertake.[32] The

26 Clause 460(1).

27 Clause 460(2).

28 Clause 461(1).

29 Clause 461(2).

30 Clause 462.

31 Clause 463(1). Residential premises are exempted from this obligation: cl 463(2).

32 Clause 464(1).

asbestos removal control plan must include details of the asbestos to be removed and the asbestos removal to be carried out, including the method to be used, tools, equipment and personal protective equipment (PPE) to be used.[33] The licensed asbestos removalist must give a copy of the asbestos removal control plan to the person who commissioned the licensed asbestos removal work.[34]

A licensed asbestos removalist must ensure that a copy of the asbestos removal control plan is kept until the asbestos removal work to which it relates is completed.[35] However, if a notifiable incident occurs in connection with the asbestos removal work, the asbestos removal control plan must be kept for at least 2 years after the incident occurs.[36] The asbestos removal control plan must be readily accessible to all PCBUs, workers, health and safety representatives, and (in relation to residential properties) the occupants of the premises.[37] It must also be available for inspection by the Regulator.[38]

A licensed asbestos removalist must give written notice to the regulator at least 5 days before the removalist commences licensed asbestos removal work.[39]

The notice must include the following:

(a) "in relation to the licensed asbestos removalist:

(i) name,

(ii) registered business name,

(iii) Australian Business Number,

(iv) licence number,

(v) business contact details,

(b) the name and business contact details of the supervisor of the licensed asbestos removal work,

(c) the name of the competent person or licensed asbestos assessor engaged to carry out a clearance inspection and issue a clearance certificate for the work,

(d) the name and contact details of the person for whom the work is to be carried out,

(e) the following in relation to the workplace where the asbestos is to be removed:

33 Clause 464(2).

34 Clause 464(3).

35 Clause 465(1).

36 Clause 465(2).

37 Clause 465(3)(a).

38 Clause 465(3)(b).

39 Clause 466(1). This obligation does not apply to emergencies where there is:"(a) a sudden and unexpected event, including a failure of equipment, that may cause persons to be exposed to respirable asbestos fibres, or (b) an unexpected breakdown of an essential service that requires immediate rectification to enable the service to continue": cl 466(2). If that is the case, the Regulator must be notified immediately by telephone and in writing within 24 hours after notice is given: cl 466(3).

Table 5.2: Selecting Respiratory Protective Equipment (Source: Safe Work Australia (2018) *Code of Practice: How to Safely Remove Asbestos*, pp 81–82)

Work procedure	Required respirator	Filter type
Simple enclosure erection for containing undamaged asbestos materials to prevent damage — no direct handling but possible disturbance of asbestos	Disposable, half-face particulate respirators OR Half-face, particulate filter (cartridge) respirator	P1 or P2
Inspection of the condition of any installed friable asbestos, which appears in poor condition or has been disturbed	Disposable, half-face particulate respirators OR Half-face, particulate filter (cartridge) respirator	P1 or P2
Sampling material for the purpose of identifying asbestos	Disposable, half-face particulate respirators OR Half-face, particulate filter (cartridge) respirator	P1 or P2
Removal of non-friable asbestos (e.g. asbestos cement sheets, ceiling tiles and vinyl tiles)	Disposable, half-face particulate respirators OR Half-face, particulate filter (cartridge) respirator	P1 or P2
Extensive sample operations on friable asbestos	Full-face, particulate filter (cartridge) respirator	P3
Maintenance work involving the removal of small quantities of friable asbestos (eg replacement of friable asbestos gaskets and insulation)	Full-face, particulate filter (cartridge) respirator	P3
Certain forms of wet stripping in which wetting is prolonged and effective, and certain small-scale dry stripping operations	Full-face, powered air-purifying particulate respirator OR Full-face, positive pressure demand air-line respirator	P3
Certain forms of dry stripping and ineffective wet stripping (light wetting, no time given to saturate)	Full-face, powered air-purifying particulate respirator OR Full-face, positive pressure demand air-line respirator No lesser respirator will suffice	P3
Dry stripping in confined areas	Full suit or hood, positive pressure demand continuous flow air-line respirator No lesser respirator will suffice	P3 only as a backup

(i) the name, including the registered business or company name, of the person with management or control of the workplace,

(ii) the address and, if the workplace is large, the specific location of the asbestos removal,

(iii) the kind of workplace,

(f) the date of the notice,

(g) the date when the asbestos removal work is to commence and the estimated duration of the work,

(h) whether the asbestos to be removed is friable or non-friable,

(i) if the asbestos to be removed is friable — the way the area of removal will be enclosed,

(j) the estimated quantity of asbestos to be removed,

(k) the number of workers who are to carry out the asbestos removal work,

(l) for each worker who is to carry out asbestos removal work — details of the worker's competency to carry out asbestos removal work".[40]

The licensed asbestos removalist must inform the person with management or control of the workplace that licensed asbestos removal work is to be carried out at the workplace and when the work is to commence.[41] In the case of residential premises, the licensed asbestos removalist is required to inform the person who commissioned the asbestos removal work, any PCBU at the workplace (such as contractors performing work on the premises), the occupier of the residential premises, the owner of the residential premises and anyone occupying premises in the immediate vicinity of the workplace.[42]

Once the person with management or control of a workplace is informed that asbestos removal work is to be carried out at the workplace they must inform their workers, any other persons at the workplace and the person who commissioned the asbestos removal work.[43] They must also take all reasonable steps to inform anyone conducting a business or undertaking at the workplace or anyone occupying premises in the immediate vicinity of the workplace.[44]

A licensed asbestos removalist must ensure that signs alerting persons to the presence of asbestos are placed to indicate where the asbestos removal work is being carried out.[45] They must also ensure that barricades are erected to delineate the asbestos removal area.[46]

The person who commissions asbestos removal work and the person with management and control of the workplace where such work is being undertaken must ensure, so far as is reasonably practicable, that no one other than workers engaged in the asbestos and other persons associated with that work (such as industrial hygienists) has access to an asbestos removal area.[47]

---

40 Clause 466(4).

41 Clause 467(2).

42 Clause 467(3).

43 Clause 468(2).

44 Clause 468(3).

45 Clause 469(a).

46 Clause 469(b).

47 Clause 470(2).

If a person has access to an asbestos removal area, the person must comply with any direction of the licensed asbestos removalist.[48] The person who commissioned the asbestos removal and the person with management or control of the workplace can refuse to allow access to an asbestos removal area at the workplace to anyone who does not comply with a control measure or a direction of the licensed asbestos removalist.[49]

A licensed asbestos removalist must ensure that facilities are available to decontaminate the following: the asbestos removal area, any plant used, workers carrying out the work and other persons who access the area.[50] They must also ensure that nothing that is likely to be contaminated with asbestos is removed from the asbestos removal area

Table 5.3: Enclosures Checklist (Source: Safe Work Australia (2018) *Code of Practice: How to Safely Remove Asbestos*, p 86)

1. Prior to dismantling the enclosure		
	Yes	No
The area within the enclosure and the area immediately surrounding the enclosure were inspected and no visible asbestos was found.	☐	☐
Air monitoring was carried out as part of the clearance inspection. The result was below 0.01 fibres/mL.	☐	☐
Is the air-monitoring report attached?	☐	☐
Can the enclosure be dismantled?	☐	☐

Number of samples collected: Click here to enter text.					
	Sample 1	Sample 2	Sample 3	Sample 4	Sample 5
Results					

2. After the enclosure was dismantled and removed		
	Yes	No
An inspection of the area in which the enclosure was erected and the area immediately surrounding the area where the enclosure was erected was inspected and no visible asbestos was found.	☐	☐
Air monitoring was carried out as part of the clearance inspection. The result was below 0.01 fibres/mL.	☐	☐
Is the air-monitoring report attached?	☐	☐
Can the area be reoccupied?	☐	☐

Number of samples collected: Click here to enter text.					
	Sample 1	Sample 2	Sample 3	Sample 4	Sample 5
Results					

48 Clause 470(5).

49 Clause 470(3).

50 Clause 471(1).

unless it is decontaminated or is sealed in a container, and the exterior of the container is, before being removed, decontaminated and labelled in accordance with the Globally Harmonised System of Classification and Labelling of Chemicals (GHS) to indicate the presence of asbestos.[51]

¶5-020 Disposal of asbestos waste

A licensed asbestos removalist must ensure that asbestos waste is contained and labelled in accordance with the GHS before the waste is removed from an asbestos removal area.[52] They must then ensure that it is disposed of as soon as practicable at a site authorised to accept asbestos waste.[53]

The Code of Practice provides that "asbestos waste must be transported and disposed of in accordance with the relevant state or territory Environment Protection Authority (EPA) requirements. Asbestos waste can only be disposed of at a site licensed by the EPA and it must never be disposed of in the general waste system".[54]

¶5-030 Disposal of PPE used in asbestos removal

PPE used in asbestos removal work and contaminated with asbestos must be sealed in a container before being removed from an asbestos waste area.[55] So far as is reasonably practicable, it must be disposed of on the completion of the asbestos removal work at a site authorised to accept asbestos waste.[56] If it is not reasonably practicable to dispose of the PPE that is clothing, the PPE must be laundered at a laundry equipped to launder asbestos-contaminated clothing.[57] If it is not practicable to launder the clothing, the PPE must be kept in the sealed container until it is re-used for asbestos removal purposes.[58] If it is not reasonably practicable to dispose of the PPE that is not clothing, the PPE must be decontaminated before it is removed from the asbestos removal area.[59] If it is not practicable to decontaminate the equipment in the asbestos removal area, the PPE must be kept in the sealed container until it is re-used for asbestos removal purposes.[60]

The licensed asbestos removalist must ensure that a sealed container is decontaminated and labelled in accordance with the GHS to indicate the presence of asbestos before being removed from the asbestos removal area.[61]

---

51 Clause 471(2).
52 Clause 472(1)(a).
53 Clause 472(1)(b).
54 Code of Practice, p 37.
55 Clause 472(2)(a).
56 Clause 472(2)(b).
57 Clause 472(2)(c)(i).
58 Clause 472(2)(c)(ii).
59 Clause 472(2)(d)(i).
60 Clause 472(2)(d)(ii).
61 Clause 272(3).

The Code of Practice further provides that "individual components and wiping rags should be placed in plastic disposal bags. Disposal bags need to be made of heavy duty plastic (200 µm polyethylene, also known as polythene) and marked with the label 'Danger Asbestos — Do not open or damage bag. Do not inhale dust'. Each bag should be sealed with adhesive (cloth or duct) tape separately prior to placing it in a second plastic asbestos disposal bag. Cloth tape is preferred when sealing asbestos waste or when sealing building enclosures as it is more durable compared to duct tape, which does not have a very long life span before it starts to shrink or fail. Asbestos waste awaiting disposal must be stored in closed containers (for example 60 or 200 l steel drums with removable lids, which should be lined with heavy duty plastic, or a sealed skip)".[62]

¶5-040 Clearance inspection

The person who commissions licensed asbestos removal work in a workplace (or if residential premises, the licensed asbestos removalist) must ensure that, when the licensed asbestos removal work is completed, a clearance inspection of the asbestos removal area at the workplace is carried out by an independent licensed asbestos assessor for work requiring a Class A asbestos removal licence or an independent competent person for all other cases.[63]

A clearance inspection is an inspection of an asbestos removal area after asbestos removal work has been completed to verify that the area is safe for normal use, which includes a visual inspection and may include air monitoring.[64]

The licensed asbestos assessor or competent person who carried out the clearance inspection must issue a clearance certificate before the asbestos removal area at the workplace is re-occupied.[65] The licensed asbestos assessor or competent person must ensure that the asbestos removal area does not pose a risk to health and safety from exposure to asbestos.[66] They must be satisfied that the asbestos removal area and the area immediately surrounding it are free from visible asbestos contamination, and if air monitoring was undertaken that the monitoring shows asbestos below 0.01 fibres/ml.[67] The clearance certificate must be in writing.[68]

62 Code of Practice, p 37.

63 Clause 473(2).

64 Clause 473(3).

65 Clause 474(2).

66 Clause 474(3).

67 Clause 474(4).

68 Clause 474(5).

¶5-050 Clearance Certificate

Table 5.4: Example of Clearance Certificate (Source: Safe Work Australia (2018) *Code of Practice: How to Safely Remove Asbestos*, p 84)

Section A — Clearance inspection details	
	Details
Client details	
Name of client:	Click here to enter text.
Client contact details:	Click here to enter text.
Removal work details	
Date removal work carried out:	Click here to enter text.
Site address where removal work is being carried out:	Click here to enter text.
Details of the specific asbestos removal work area(s):	Click here to enter text.
Name of licensed asbestos removalist:	Click here to enter text.
Name and contact details of licensed asbestos removalist supervisor (if different to removalist):	Click here to enter text.
Inspection details	
Date of clearance inspection:	Click here to enter a date.
Time of clearance inspection:	Click here to enter text.

Section B — Asbestos removal work paperwork		
	Yes	No
Do you have a copy of the asbestos removal control plan?	☐	☐
Do you have a copy of the notification form?	☐	☐
Is the removal work (eg use of enclosures, decontamination facilities, waste facilities) consistent with the control plan and the notification form?	☐	☐

Section C — Asbestos removal work area		
1. Visual inspection		
	Yes	No
Inspection of the specific area detailed in Section A found no visible asbestos remaining as a result of the asbestos removal work carried out.	☐	☐
Is air monitoring required? (if no, proceed to Section E)	☐	☐
Can the area be reoccupied?	☐	☐
Has additional information been attached? (eg photos, drawings, plans)	☐	☐

2. Air monitoring		
	Yes	No
Air monitoring was carried out as part of the clearance inspection. The result was below 0.01 fibres/mL.	☐	☐
Has the air-monitoring sample been analysed by a NATA-accredited laboratory or a laboratory approved by the WHS regulator?	☐	☐
Is the air-monitoring report attached?	☐	☐
Can the area be reoccupied?	☐	☐

¶5-060 Air monitoring for Class A asbestos removal work

The PCBU who commissions asbestos removal work requiring a Class A asbestos removal licence at a workplace must ensure that an independent licensed asbestos assessor undertakes air monitoring of the asbestos removal area at the workplace.[69] If the workplace is residential premises, that obligation falls on the licensed asbestos removalist carrying out asbestos removal work.[70]

The air monitoring must be carried out immediately before the licensed asbestos removal work commences (unless glove bags are to be used for the removal) and while the licensed asbestos removal work is carried out.[71]

The person who commissions the licensed asbestos removal work must ensure that the results of the air monitoring are given to workers, health and safety representatives, PCBUs and other persons at the workplace.[72] If the workplace is residential premises, that obligation falls on the licensed asbestos removalist carrying out the work.[73]

The independent licensed asbestos assessor, who undertakes air monitoring must use the membrane filter method for the air monitoring.[74]

¶5-070 Action if respirable asbestos fibre level too high

If respirable asbestos fibre levels are recorded at the asbestos removal area at 0.01 fibres/ml or more, but not more than 0.02 fibres/ml, the licensed asbestos removalist must immediately:

- investigate the cause of the respirable asbestos fibre level
- implement controls to prevent exposure of anyone to asbestos, and
- prevent the further release of respirable asbestos fibres.[75]

---

69 Clause 475(1).

70 Clause 475(2).

71 Clause 475(3).

72 Clause 475(4).

73 Clause 475(5).

74 Clause 475(6).

75 Clause 476(1)(a).

If respirable asbestos fibre levels are recorded at the asbestos removal area at more than 0.02 fibres/ml, they must immediately:

- order the asbestos removal work to stop
- notify the regulator
- investigate the cause of the respirable asbestos fibre level
- implement controls to prevent exposure of anyone to asbestos, and
- prevent the further release of respirable asbestos fibre.[76]

If the licensed removalist stops asbestos removal work requiring a Class A asbestos removal licence because the recorded respirable asbestos fibre level exceeds 0.02 fibres/ml, the removalist must ensure that the asbestos removal work does not resume until air monitoring shows that the recorded respirable asbestos fibre level is below 0.01 fibres/ml.[77]

¶5-080 Removing friable asbestos

The licensed asbestos removalist removing friable asbestos must ensure, so far as is reasonably practicable, that the asbestos removal area is enclosed to prevent the release of respirable asbestos fibres.[78] They must ensure that negative pressure is used.[79] They must ensure that the wet method of asbestos removal is used.[80] They must also ensure that the asbestos removal work does not commence until the air monitoring is commenced by a licensed asbestos assessor.[81] The air monitoring must be undertaken during the asbestos removal work, at times decided by the independent licensed asbestos assessor undertaking the monitoring.[82] Any glove bag used to enclose the asbestos removal area must be dismantled and disposed of safely.[83]

The licensed asbestos removalist must ensure that any enclosure used in removing friable asbestos is tested for leaks.[84]

The licensed removalist must not dismantle an enclosure for a friable asbestos removal area until the removalist receives results of air monitoring, showing that the recorded respirable asbestos fibre level within the enclosure is below 0.01 fibres/ml, from the person who commissioned the work (or if residential premises, from the licensed asbestos assessor who undertook the air monitoring).[85]

---

76 Clause 476(1)(b).

77 Clause 476(2).

78 Clause 477(1)(a).

79 Clause 477(1)(b). This obligations does not apply if glove bags are used.

80 Clause 477(1)(c).

81 Clause 477(1)(d). This obligations does not apply if glove bags are used.

82 Clause 477(1)(e).

83 Clause 477(1)(f).

84 Clause 477(2).

85 Clause 477(4).

The licensed removalist must ensure that an enclosure for a friable asbestos removal area is dismantled in a way that, so far as is reasonably practicable, eliminates the release of respirable asbestos fibre.[86]

The person who commissioned the removal of the friable asbestos must obtain a clearance certificate from a licensed asbestos assessor after the enclosure for the friable asbestos removal area has been dismantled.[87]

86 Clause 477(5).

87 Clause 477(6).

CHAPTER 6

HEALTH MONITORING

A person conducting business or undertaking (PCBU) must ensure that health monitoring is provided to a worker carrying out work for the business or undertaking if the worker is carrying out licensed asbestos removal work at a workplace and is at risk of exposure to asbestos when carrying out the work.[1] They are also required to provide health monitoring to a worker who is carrying out other ongoing asbestos removal work or asbestos-related work and is at risk of exposure to asbestos when carrying out the work.[2] The health monitoring of the worker must commence before the worker carries out licensed asbestos removal work.[3] The worker must be informed of any health monitoring requirements before carrying out any work that may expose the worker to asbestos.[4]

The Code of Practice provides that the "need for health monitoring for workers at risk of exposure to asbestos should be determined on the basis of:

- the potential for exposure
- the frequency of potential exposure, and
- the duration of the work being undertaken".[5]

The Code of Practice further provides that health monitoring "should also be provided to the worker at regular intervals (at least once every 2 years) after the worker commences the asbestos-related work".[6]

Unless otherwise recommended by a registered medical practitioner, the health monitoring must include consideration of the worker's demographic, medical and occupational history and records of the worker's personal exposure as well as a physical examination of the worker.[7]

1 Clause 435(1)(a).

2 Clause 435(1)(b).

3 Clause 435(2).

4 Clause 435(3).

5 Code of Practice, p 39.

6 Ibid.

7 Clause 436.

The health monitoring must be carried out by or under the supervision of a registered medical practitioner with experience in health monitoring.[8] The worker must be consulted in relation to the selection of the registered medical practitioner.[9]

The PCBU must provide the following information to the registered medical practitioner carrying out or supervising the health monitoring:

- the name and address of the person conducting the business or undertaking
- the name and date of birth of the worker
- the work that the worker is, or will be, carrying out that has triggered the requirement for health monitoring, and
- if the worker has started that work, how long the worker has been carrying out that work.[10]

The cost of the health monitoring must be met by the PCBU.[11] If two or more PCBUs have a duty to provide health monitoring for a worker and have arranged for one of them to commission the health monitoring, the costs of the health monitoring for which any of those persons is liable must be apportioned equally between each of those persons unless they agree otherwise.[12]

The PCBU must take all reasonable steps to obtain a health monitoring report from the registered medical practitioner who carried out or supervised the monitoring as soon as practicable after the monitoring is carried out in relation to a worker.[13] The health monitoring report must include the following:

- the name and date of birth of the worker
- the name and registration number of the registered medical practitioner
- the name and address of the person conducting the business or undertaking who commissioned the health monitoring
- the date of health monitoring
- any advice that test results indicate that the worker may have contracted a disease, injury or illness as a result of carrying out the work that triggered the requirement for health monitoring
- any recommendation that the person conducting the business or undertaking takes remedial measures, including whether the worker can continue to carry out the type of work that triggered the requirement for health monitoring, and

8 Clause 437(1).
9 Clause 437(2).
10 Clause 439.
11 Clause 438(1).
12 Clause 438(2).
13 Clause 440(1).

- whether medical counselling is required for the worker in relation to the work that triggered the requirement for health monitoring.[14]

The report must be provided to the relevant worker as soon as practicable after the PCBU obtains the report.[15] It must also be provided to the Regulator as soon as practicable after obtaining the report if the report contains any advice that test results indicate that the worker may have contracted a disease, injury or illness or any recommendation that the person conducting the business or undertaking takes remedial measures.[16]

The PCBU who commissioned health monitoring for a worker must give a copy of the health monitoring report to all other persons conducting businesses or undertakings who have a duty to provide health monitoring for the worker as soon as practicable after obtaining the report.[17]

Health monitoring reports in relation to a worker must be kept as a confidential record, identified as a record in relation to the relevant worker[18]and be kept for at least 40 years after the record is made.[19]

The PCBU must ensure that the health monitoring report and results of a worker are not disclosed to another person without the worker's written consent.[20]

14 Clause 440(2).

15 Clause 441.

16 Clause 442.

17 Clause 443.

18 Clause 444(1)(a).

19 Clause 444(1)(b).

20 Clause 444(2).

CHAPTER 7

THE WAY FORWARD

¶7-010 It's about time we get serious about asbestos eradication

We have known about the adverse health effects of asbestos from the very outset of its manufacture. Early warning of the health hazard of asbestos was sounded in England in the late 19th century almost immediately after the start of asbestos manufacture.[1] One of the inspectors of factories described asbestos as being a dust easily demonstrated as a danger to health.[2]

In 1900, an autopsy revealed that an asbestos worker had suffered from pulmonary fibrosis most likely to have been caused by his exposure to asbestos.[3] The patient declared that he was the last survivor of a group of 10 workers who started together.[4]

In 1924, the first detailed account of a case of asbestosis was published.[5] Scientific interest was raised, and within a few years, several pathologists and physicians published case reports of pulmonary fibrosis in asbestos workers.[6] The term "pulmonary asbestosis", first introduced in 1927,[7] became widely used in the medical

1 Burdorf, L, Swuste, P and Heederik, D (1991) "A history of awareness of asbestos disease and the control of occupational asbestos exposures in The Netherlands", *American Journal of Industrial Medicine* 20:547–555 (1991) at 548.

2 Ibid.

3 Montague-Murray, M (1907), "Statement before the Committee in the Minutes of Evidence", in "Report of the Departmental Committee on Compensation for Industrial Diseases.", London: His Majesty's Stationery Office, cited in Burdorf et al (1991) op cit 208.

4 Ibid.

5 Cooke, W (1924), "Fibrosis of the lungs due to inhalation of asbestos dust", *British Medical Journal* 11:147–148 cited in Burdorf et al (1991) op cit 208.

6 McDonald, S (1927), "Histology of pulmonary asbestosis", *British Medical Journal*, 11:1024–1025 cited in Burdorf et al (1991) op cit 208. See also Oliver, T (1927), "Clinical aspects of pulmonary asbestosis", *British Medical Journal*, 11:1026–1027 and Simson, F (1928), "Pulmonary asbestosis in South Africa", *British Medical Journal*, 1:885–887.

7 Cooke, W (1927), "Pulmonary asbestosis", *British Medical Journal*, 11:1024 cited in Burdorf et al (1991) op cit 208.

literature.[8] One of the first epidemiologic surveys, carried out by the British Factory Inspectorate, provided convincing evidence of the relationship between occupational exposure to asbestos and the development of pulmonary fibrosis.[9] Indeed, Asbestosis was officially recognised as a compensable occupational disease in the United Kingdom in 1931.[10] Yet, it was not until 31 December 2003 that asbestos was finally banned in Australia. Even now, unless asbestos is disturbed, there is no express legal obligation to remove it. Indeed, until the mid-1980s, Australia had one of the highest per capita rates of asbestos use in the world with 1.5 million tonnes of asbestos imported into Australia between 1930 and 1983 when the health effects of asbestos use was well known.[11] The problem is so pervasive in Australia that it is estimated that one-third of all Australian homes contain asbestos products.[12]

This is against the background of apathy for asbestos exposure among the community. Douglas and Van der Borre (2019) argue that the risks from asbestos exposure remain poorly understood, even in high-risk groups.[13] A survey commissioned by the Australian Government agency established to raise awareness of asbestos safety, the Asbestos Eradication Agency (AEA) supports this assertion.

In 2014, AEA commissioned a national survey of awareness of and attitudes towards asbestos and the risks of exposure to asbestos. The survey was intended as a baseline of statistics on which the agency could test awareness levels and attitudes at biennial intervals into the future. A further survey was conducted in 2016. Surprisingly, the results indicated that awareness levels have plateaued and in some places have gone backwards.[14]

The time has come for our attitudes towards asbestos to change. There should be an express obligation to remove asbestos when it is identified. In this book, I have argued that, in a workplace context, this is the effect of the duty of care and the duty of officers. No risk assessment is complete if it does not provide due consideration to eradication of asbestos. Given the health effects of asbestos, it is difficult to see how an assessment of reasonable practicability steps can progress past elimination as the appropriate control to the identification of asbestos. As considered, long-term cost benefit analysis would support that conclusion.

---

8 See Haddow, A (1929) "Clinical aspects of pulmonary asbestosis", *British Medical Journal*, 11:580–581; Stewart, M (1929) "A method for examining the sputum for asbestos bodies" *British Medical Journal*, 11:581; and Wood, W (1929), "Pulmonary asbestosis. Radiographic appearance in skiagrams of the chest of workers in asbestos", Tubercle 10:353–363 cited in Burdorf et al (1991) op cit 208.

9 Merewether, E and Price, C (1930), "Report on the Effects of Asbestos Dust on the Lungs and Dust Suppression in the Asbestos Industry", London, His Majesty's Stationery Office, cited in Burdorf et al (1991) op cit 208.

10 Burdorf et al (1991) op cit 208. The link between asbestos and Mesothelioma was known since the 1950s.

11 Asbestos Eradication Agency, "Asbestos use if Australia", accessed online at www.asbestossafety.gov.au. Asbestos was mined in Australia until 1984.

12 Ibid.

13 Douglas, T and Van der Borre, L (2019), "Asbestos neglect: Why asbestos exposure deserves greater policy attention", *Health Policy*, 123(5): 516–519.

14 Asbestos Eradication Agency, "Trends in asbestos awareness and management", accessed online at www. asbetossafety.gov.au.

Index

Made in the USA
Monee, IL
05 November 2026

67811058R00046